# LINEAR PROGRAMMING

# LIBRARY OF MATHEMATICS

## edited by

### WALTER LEDERMANN

D.Sc., Ph.D., F.R.S.Ed., Professor of
Mathematics, University of Sussex

# LINEAR PROGRAMMING

BY

## KATHLEEN TRUSTRUM

LONDON: ROUTLEDGE & KEGAN PAUL LTD.

First published 1971
in Great Britain by
Routledge & Kegan Paul Ltd
Broadway House, 68–74 Carter Lane
London, EC4 SEL
© Kathleen Trustrum 1971
Printed in Hungary

ISBN 0 7100 6972 3 (c)
ISBN 0 7100 6779 8 (p)

# Contents

# Preface

Linear programming is a relatively modern branch of Mathematics, which is a result of the more scientific approach to management and planning of the post-war era. The purpose of this book is to present a mathematical theory of the subject, whilst emphasising the applications and the techniques of solution. An introduction to the theory of games is given in chapter five and the relationship between matrix games and linear programmes is established.

The book assumes that the reader is familiar with matrix algebra and the background knowledge required is covered in the book, *Linear Equations* by P.M. Cohn, of this series. In fact the notation used in this text conforms with that introduced by Cohn.

The book is based on a course of about 18 lectures given to Mathematics and Physics undergraduates. Several examples are worked out in the text and each chapter is followed by a set of examples.

I am grateful to my husband for many valuable suggestions and advice, and also to Professor W. Ledermann, for encouraging me to write this book.

*University of Sussex*                                    Kathleen Trustrum

# CHAPTER ONE

# Convex Sets

## 1. Convex Hulls, Polytopes and Vertices

The theorems proved in this chapter provide the mathematical background for the theory of linear programming. Convex sets are introduced, not only to prove some of the theorems, but also to give the reader a geometrical picture of the algebraic processes involved in solving a linear programme. In the following definitions, $R^n$ denotes the vector space of all real n-vectors.

DEFINITION. A subset[1] $C \subset R^n$ is convex if for all[2] $u_1, u_2 \in C$.

$$\lambda u_1 + (1 - \lambda)u_2 \in C \text{ for } 0 \leq \lambda \leq 1.$$

In other words, a set is convex if the straight line segment joining any two points of the set, also belongs to the set. For example, a sphere and a tetrahedron are convex sets, whereas a torus is not.

DEFINITIONS. A vector $x$ is a convex combination of $u_1, u_2, \ldots, u_k \in R^n$ if

$$x = \sum_{i=1}^{k} \lambda_i u_i, \quad \text{where } \lambda_i \geq 0 \text{ and } \sum_{i=1}^{k} \lambda_i = 1.$$

The convex hull of a set $X \subset R^n$, written $\langle X \rangle$, is the set of all convex combinations of points of $X$. If $X$ is a finite set, then the convex hull is called a convex polytope.

---

[1] $\subset$ is contained in.
[2] $\in$ belongs to.

It is easily shown that the convex hull of a set is a convex set. In two dimensions a convex polytope is simply a convex polygon and figure 1 shows the convex hull $\langle X \rangle$ of the finite set of points $X = \{\mathbf{u}_1, \mathbf{u}_2, \ldots, \mathbf{u}_6\} \subset R^2$, which is an example of a convex polytope.

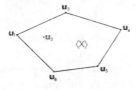

FIGURE 1

A vertex or extreme point of a convex set $C$ is a point belonging to $C$, which is not the interior point of any straight line segment of $C$. The vertices of the convex set shown in figure 1 are $\mathbf{u}_1$, $\mathbf{u}_3$, $\mathbf{u}_4$, $\mathbf{u}_5$ and $\mathbf{u}_6$. A convex set may have no vertices e.g. the interior of a circle. A more precise definition of a vertex is now given.

DEFINITION. A vertex $\mathbf{x}$ of a convex set $C$ belongs to $C$ and is such[1] that, for any $\mathbf{y}$, $\mathbf{z}$ belonging to $C$ and $0 < \lambda < 1$,

$$\mathbf{x} = \lambda\mathbf{y} + (1 - \lambda)\mathbf{z} \Rightarrow \mathbf{x} = \mathbf{y} = \mathbf{z}.$$

The first theorem proves that a convex polytope is the convex hull of its vertices (see figure 1).

THEOREM 1. If $C$ is a convex polytope and $V$ is its set of vertices then

$$C = \langle V \rangle$$

*Proof.* Since $C$ is a convex polytope, $C = \langle \mathbf{u}_1, \mathbf{u}_2, \ldots, \mathbf{u}_k \rangle$. A minimal subset $\{\mathbf{v}_1, \mathbf{v}_2, \ldots, \mathbf{v}_r\}$ is selected from amongst the $\mathbf{u}_i$'s so that $C = \langle \mathbf{v}_1, \mathbf{v}_2, \ldots, \mathbf{v}_r \rangle$, where $r \leq k$. This is

[1] $\Rightarrow$ implies.

achieved by eliminating any $\mathbf{u}_j$ which is a convex combination of the other $\mathbf{u}_i$'s. To show that $\mathbf{v}_1$ is a vertex, we suppose that

$$\mathbf{v}_1 = \lambda\mathbf{x} + (1 - \lambda)\mathbf{y}, \text{ where } \mathbf{x}, \mathbf{y} \in C \text{ and } 0 < \lambda < 1.$$

Expressing $\mathbf{x}$ and $\mathbf{y}$ as a convex combination of the $\mathbf{v}_i$'s, we have that $\mathbf{v}_1 = \lambda \sum_{i=1}^{r} \lambda_i \mathbf{v}_i + (1 - \lambda) \sum_{i=1}^{r} \mu_i \mathbf{v}_i$, where $\sum_{i=1}^{r} \lambda_i = \sum_{i=1}^{r} \mu_i = 1$, $\lambda_i \geq 0$, $\mu_i \geq 0$.
Hence

$$(1 - \alpha_1)\mathbf{v}_1 = \sum_{i=2}^{r} \alpha_i \mathbf{v}_i, \text{ where } \alpha_i = \lambda\lambda_i + (1 - \lambda)\mu_i \geq 0 \text{ and}$$

$$\sum_{i=1}^{r} \alpha_i = 1. \tag{1}$$

If $\alpha_1 < 1$, (1) implies that $\mathbf{v}_1$ is a convex combination of $\mathbf{v}_2$, ..., $\mathbf{v}_r$, which is false, so $\alpha_1 = 1$ and since $0 < \lambda < 1$, $\lambda_1 = \mu_1 = 1$. It now follows that $\mathbf{v}_1 = \mathbf{x} = \mathbf{y}$ and $\mathbf{v}_1$ is a vertex. Similarly $\mathbf{v}_2, \ldots, \mathbf{v}_r$, are vertices and it is clear that $C$ has no other vertices, hence $V = \{\mathbf{v}_1, \mathbf{v}_2, \ldots, \mathbf{v}_r\}$ and $C = \langle V \rangle$.

We now use Theorem 1 to prove that a linear function defined on a convex polytope $C$ attains its maximum (minimum) in $C$ at a vertex of $C$. The general linear programming problem (see page 17) is to maximise or minimise a linear function subject to linear constraints. Such constraints define a convex set and in the cases where the convex set is also a polytope, the maximum and minimum can be found by evaluating the linear function at the vertices.

THEOREM 2. If $C \subset R^n$ is a convex polytope, then the linear function[1] $\mathbf{c}'\mathbf{x}$ takes its maximum (minimum) in $C$ at a vertex of $C$, where constant $\mathbf{c}$ and variable $\mathbf{x}$ belong to $R^n$.

[1] $\mathbf{x} = \begin{pmatrix} x_1 \\ x_2 \\ \vdots \\ x_n \end{pmatrix}$ is a column vector, $\mathbf{c}' = (c_1, c_2, \ldots, c_n)$ is a row vector and $\mathbf{c}'\mathbf{x} = c_1 x_1 + c_2 x_2 + \ldots + c_n x_n$.

*Proof.* Since $C$ is a convex polytope, it follows from Theorem 1 that $C = \langle \mathbf{v}_1, \mathbf{v}_2, \ldots, \mathbf{v}_r \rangle$, where $\mathbf{v}_i$ are the vertices of $C$. Let $M = \max_i \mathbf{c}' \mathbf{v}_i$, then for any $\mathbf{x} \in C$

$$\mathbf{x} = \sum_{i=1}^{r} \lambda_i \mathbf{v}_i, \text{ where } \lambda_i \geq 0 \text{ and } \sum_{i=1}^{r} \lambda_i = 1$$

$$\text{and} \quad \mathbf{c}' \mathbf{x} = \sum_{i=1}^{r} \lambda_i \mathbf{c}' \mathbf{v}_i \leq M \sum_{i=1}^{r} \lambda_i = M .$$

Hence the maximum in $C$ is attained at a vertex of $C$.

## 2. Basic Solutions of Equations

Often the convex set $C$, on which a linear function is to be maximised or minimised, consists of the non-negative solutions of a system of linear equations, that is

$$C = \{\mathbf{x} \mid A\mathbf{x} = \mathbf{b}, \ \mathbf{x} \geq \mathbf{0}\},[1]$$

where $A$ is an $m \times n$ matrix, $\mathbf{x}$ is an $n$-vector and $\mathbf{b}$ is an $m$-vector. To prove that $C$ is convex, let $\mathbf{u}_1, \mathbf{u}_2 \in C$ then

$$A (\lambda \mathbf{u}_1 + (1 - \lambda)\mathbf{u}_2) = \lambda \mathbf{b} + (1 - \lambda)\mathbf{b} = \mathbf{b}$$

and $\quad \lambda \mathbf{u}_1 + (1 - \lambda)\mathbf{u}_2 \geq \mathbf{0} \quad$ for $0 \leq \lambda \leq 1$

so $\quad \lambda \mathbf{u}_1 + (1 - \lambda)\mathbf{u}_2 \in C \quad$ for $0 \leq \lambda \leq 1$.

An alternative representation of the system of equations $A\mathbf{x} = \mathbf{b}$ is $\sum_{i=1}^{n} x_i \mathbf{a}_i = \mathbf{b}$, where $\mathbf{a}_i$ are the column vectors of $A$.

---

[1] $\{\mathbf{x} \mid p\}$ the set of all $\mathbf{x}$ with the property $p$.

4

DEFINITION. A solution of the system of equations

$$\sum_{i=1}^{n} x_i \mathbf{a}_i = \mathbf{b} \tag{2}$$

is called basic, if the set of vectors $\{\mathbf{a}_i \mid x_i \neq 0\}$ is linearly independent.

We now identify the basic non-negative solutions of (2) with the vertices of $C = \{\mathbf{x} \mid A\mathbf{x} = \mathbf{b}, \mathbf{x} \geq \mathbf{0}\}$. Let $\mathbf{x} \geq \mathbf{0}$ be a basic solution of (2), then $\mathbf{x} \in C$. Suppose that

$$\mathbf{x} = \lambda\mathbf{y} + (1 - \lambda)\mathbf{z}, \quad \text{where} \quad 0 < \lambda < 1 \quad \text{and} \quad \mathbf{y}, \mathbf{z} \in C. \tag{3}$$

Since $\mathbf{x} \geq \mathbf{0}$, $\mathbf{y} \geq \mathbf{0}$, $\mathbf{z} \geq \mathbf{0}$ and $0 < \lambda < 1$,

$$x_i = 0 \Longrightarrow y_i = z_i = 0. \tag{4}$$

Also $\mathbf{y}$ and $\mathbf{z}$ satisfy (2), therefore

$$\sum_{i \in S} (y_i - z_i)\mathbf{a}_i = \mathbf{0}, \quad \text{where} \quad S = \{i \mid x_i > 0\}.$$

Since $\mathbf{x}$ is basic, the set of vectors $\{\mathbf{a}_i \mid i \in S\}$ is linearly independent,

$$\text{so } y_i = z_i \text{ for } i \in S. \tag{5}$$

It now follows from (3), (4) and (5) that $\mathbf{x} = \mathbf{y} = \mathbf{z}$, so $\mathbf{x}$ is a vertex of $C$. We leave the reader to prove that a vertex of $C$ is a basic non-negative solution of (2) (see exercise 4 on page 12).

The following theorem is important and proves that if a system of equations has a non-negative solution, then it has a basic non-negative solution. This is equivalent to saying that if $C$ is non-empty, then $C$ has at least one vertex.

THEOREM 3. If the system of equations $\sum_{i=1}^{n} x_i\mathbf{a}_i = \mathbf{b}$ has a non-negative solution, then it has a basic non-negative solution.

*Proof.* The theorem is proved by induction on $n$. For $n = 1$ the result is immediate and the inductive assumption is that the result holds for the system of equations

$$\sum_{i=1}^{k} x_i \mathbf{a}_i = \mathbf{b} \tag{6}$$

for $k < n$. Let $\mathbf{x} = (x_1, \ldots, x_n)' \geq \mathbf{0}$ be a solution of (6) for $k = n$. If $\mathbf{x}$ is basic, there is nothing to prove and if any $x_i = 0$, we are reduced to the case $k < n$. We are now left with the case in which $\mathbf{x}$ is non-basic and $x_i > 0$ for all $i$. Since $\mathbf{x}$ is non-basic, $\mathbf{a}_1, \mathbf{a}_2, \ldots, \mathbf{a}_n$ are linearly independent and there exist $\lambda_i$, not all zero, such that

$$\sum_{i=1}^{n} \lambda_i \mathbf{a}_i = \mathbf{0}. \tag{7}$$

Without loss of generality some $\lambda_i > 0$, otherwise (7) can be multiplied by $-1$. Choose

$$\theta = \min_{\lambda_i > 0} \frac{x_i}{\lambda_i} = \frac{x_n}{\lambda_n} \qquad \text{(say)}.$$

It now follows from (6) and (7) that

$$\sum_{i=1}^{n} (x_i - \theta \lambda_i) \mathbf{a}_i = \mathbf{b}, \text{ where } x_i - \theta \lambda_i \geq 0 \text{ for all } i.$$

Since $x_n - \theta \lambda_n = 0$, $(x_1 - \theta \lambda_1, x_2 - \theta \lambda_2, \ldots, x_{n-1} - \theta \lambda_{n-1})$ is a non-negative solution of (6) with $k = n - 1$, and so by the inductive assumption, the equations have a basic non-negative solution.

The following example illustrates some of the preceding results and shows that $C = \{\mathbf{x} \mid A\mathbf{x} = \mathbf{b}, \mathbf{x} \geq \mathbf{0}\}$ is not necessarily a convex polytope.

*Example.* Determine the nature of the convex set $C$ of the non-negative solutions of the equations,

$$\alpha x_1 + x_2 = 3, \qquad x_2 - x_3 = \alpha \tag{8}$$

for all values of $\alpha$.

*Solution.* The basic solutions of (8) are found by putting $x_1$, $x_2$ and $x_3$, equal to zero in turn, which gives

$$\mathbf{u}_1 = (0, 3, 3 - \alpha)' \geq \mathbf{0} \qquad \text{for } \alpha \leq 3,$$

$$\mathbf{u}_2 = (3\alpha^{-1}, 0, -\alpha)' \not\geq \mathbf{0},$$

$$\mathbf{u}_3 = (\alpha^{-1}(3 - \alpha), \alpha, 0)' \geq \mathbf{0} \quad \text{for } 0 < \alpha \leq 3.$$

Since the basic non-negative solutions of (8) are the vertices of $C$, $C$ has no vertices for $\alpha > 3$ and so by Theorem 3, $C$ is empty. For $\alpha = 3$, $C$ has one vertex $\mathbf{u}_1(= \mathbf{u}_3)$, for $0 < \alpha < 3$, $C$ has two vertices $\mathbf{u}_1$ and $\mathbf{u}_3$, and for $\alpha \leq 0$, $C$ has one vertex $\mathbf{u}_1$.

The general solution of (8) is

$$\mathbf{x} = (x_1, 3 - \alpha x_1, \ 3 - \alpha x_1 - \alpha)' = \mathbf{u}_1 + x_1(1, -\alpha, -\alpha)',$$

which represents a line through the point $\mathbf{u}_1$. The condition $\mathbf{x} \geq \mathbf{0}$ implies that $C = \mathbf{u}_1$ for $\alpha = 3$, $C$ is the convex polytope $\langle \mathbf{u}_1, \mathbf{u}_3 \rangle$ for $0 < \alpha < 3$ and $C$ is the half-line with vertex $\mathbf{u}_1$ for $\alpha \leq 0$, which is not a convex polytope. Geometrically $C$ is the intersection of a line with the positive octant, which may be empty, a single point, a finite line segment or a half-line. These possibilities are illustrated in figure 2.

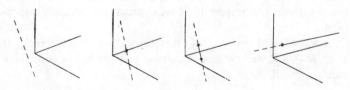

FIGURE 2

### 3. Theorem of the Separating Hyperplane

We now leave linear equations temporarily and prove a more general result about convex sets, which is known as the theorem of the separating hyperplane. This theorem asserts that if a point does not belong to a closed convex set in $R^n$, then a hyperplane (an $(n-1)$-dimensional plane) can be drawn so that the point and convex set lie on opposite sides of the hyperplane (see figure 3). The theorem does not hold if the set is not closed for then the point can be chosen as one of the limit points

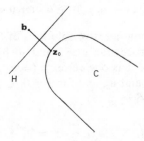

FIGURE 3

not belonging to the set, in which case it is impossible to draw a separating hyperplane. For example, take the convex set consisting of the interior of a circle in $R^2$ and take any point on the circumference of the circle. Nor does the theorem hold if the set is not convex, for then there is a line segment joining two points of the set, which is not wholly contained in the set. Choose a point on this line segment which does not belong to the set, then it is impossible to construct a separating hyperplane.

THEOREM 4 (theorem of the separating hyperplane). If $C \subset R^n$ is a closed convex set and the $n$-vector $\mathbf{b} \notin C$, then there exists a $\mathbf{y} \in R^n$ such that

$$\mathbf{y}'\mathbf{b} > \mathbf{y}'\mathbf{z} \text{ for all } \mathbf{z} \in C.$$

*Proof.* The proof is analytical. Choose a closed sphere $S$ centre $\mathbf{b}$ such that $C \cap S$ is non-empty, then the set $C \cap S$ is compact[1]. Since the function $f(\mathbf{z}) = (\mathbf{b} - \mathbf{z})'(\mathbf{b} - \mathbf{z})$ is continuous, it is bounded and attains its bounds in $C \cap S$, that is there exists a point $\mathbf{z}_0 \in C \cap S$ such that[2]

$$(\mathbf{b} - \mathbf{z}_0)'(\mathbf{b} - \mathbf{z}_0) = \underset{\mathbf{z} \in C \cap S}{\text{g.l.b.}} (\mathbf{b} - \mathbf{z})'(\mathbf{b} - \mathbf{z}).$$

The above result clearly holds for all $\mathbf{z} \in C$ and since $\mathbf{b}$ and $\mathbf{z}_0$ are distinct points,

$$0 < (\mathbf{b} - \mathbf{z}_0)'(\mathbf{b} - \mathbf{z}_0) \leq (\mathbf{b} - \mathbf{z})'(\mathbf{b} - \mathbf{z}) \quad \text{for all} \quad \mathbf{z} \in C. \quad (9)$$

It follows from the convexity of $C$ that for all $\mathbf{z} \in C$ and $0 \leq \lambda \leq 1$

$$(\mathbf{b} - \mathbf{z}_0)'(\mathbf{b} - \mathbf{z}_0) \leq (\mathbf{b} - \lambda\mathbf{z} - (1 - \lambda)\mathbf{z}_0)'(\mathbf{b} - \lambda\mathbf{z} - (1 - \lambda)\mathbf{z}_0)$$

which can be written as

$$0 \leq \lambda^2(\mathbf{z}_0 - \mathbf{z})'(\mathbf{z}_0 - \mathbf{z}) + 2\lambda(\mathbf{b} - \mathbf{z}_0)'(\mathbf{z}_0 - \mathbf{z}).$$

For the above inequality to hold for all $\lambda$ satisfying $0 \leq \lambda \leq 1$,

$$(\mathbf{b} - \mathbf{z}_0)'(\mathbf{z}_0 - \mathbf{z}) \geq 0 \quad \text{for all} \quad \mathbf{z} \in C.$$

Combining the left-hand side of (9) with the last inequality gives

$$(\mathbf{b} - \mathbf{z}_0)'\mathbf{b} > (\mathbf{b} - \mathbf{z}_0)'\mathbf{z}_0 \geq (\mathbf{b} - \mathbf{z}_0)'\mathbf{z} \quad \text{for all} \quad \mathbf{z} \in C,$$

and on putting $\mathbf{y} = (\mathbf{b} - \mathbf{z}_0)$, the required result,

$$\mathbf{y}'\mathbf{b} > \mathbf{y}'\mathbf{z} \quad \text{for all} \quad \mathbf{z} \in C,$$

follows. (A separating hyperplane is $H = \{\mathbf{x} \mid \mathbf{y}'\mathbf{x} = \frac{1}{2} \mathbf{y}'(\mathbf{z}_0 + \mathbf{b})\}$. See figure 3.)

---

[1] $\cap$ intersection.
[2] g. l. b. greatest lower bound or inf.

## 4. Alternative Solutions of Linear Inequalities

We now use the theorem of the separating hyperplane in the proof of the next theorem.

THEOREM 5. Either

$$\text{the equations } A\mathbf{x} = \mathbf{b} \text{ have a solution } \mathbf{x} \geq \mathbf{0}, \quad (10)$$

or

$$\text{the inequalities } \mathbf{y}'A \leq \mathbf{0}, \ \mathbf{y}'\mathbf{b} > 0 \text{ have a solution}, \quad (11)$$

but not both.

*Proof.* If both (10) and (11) hold, then

$$0 \geq \mathbf{y}'A\mathbf{x} = \mathbf{y}'\mathbf{b} > 0,$$

which is impossible. Suppose (10) is false, then

$$\mathbf{b} \notin C = \{\mathbf{z} = A\mathbf{x} \mid \mathbf{x} \geq \mathbf{0}\}.$$

It is easy to show that $C$ is convex and it can be proved that $C$ is closed, so by Theorem 4, there exists a $\mathbf{y}$ satisfying

$$\mathbf{y}'\mathbf{b} > \mathbf{y}'\mathbf{z} = \mathbf{y}'A\mathbf{x} \text{ for all } \mathbf{x} \geq \mathbf{0}. \quad (12)$$

In particular (12) is satisfied for $\mathbf{x} = \mathbf{0}$, so $\mathbf{y}'\mathbf{b} > 0$. To prove that $\mathbf{y}'A \leq \mathbf{0}$, suppose for contradiction that

$$(\mathbf{y}'A)_i = \lambda > 0,$$

then for $\mathbf{x} = \lambda^{-1}(\mathbf{y}'\mathbf{b})\mathbf{e}_i \geq \mathbf{0}$, where $\mathbf{e}_i$ is the $i$th unit vector,[1]

$$\mathbf{y}'A\mathbf{x} = \lambda^{-1}(\mathbf{y}'\mathbf{b})(\mathbf{y}'A)_i = \mathbf{y}'\mathbf{b},$$

which contradicts (12). Hence the inequalities in (11) have a solution, so (10) and (11) cannot be simultaneously false, which establishes the theorem.

[1] $\mathbf{e}_i$ the vector whose $i$th component is one and whose other components are zero, $I = \begin{pmatrix} 1 & 0 & \dots & 0 \\ 0 & 1 & \dots & 0 \\ \vdots & \vdots & & \vdots \\ 0 & 0 & & 1 \end{pmatrix}$

The following theorem is a corollary to Theorem 5 and will be used later in the proof of the fundamental duality theorem of Linear Programming.

THEOREM 6. Either

the inequalities $A\mathbf{x} \geq \mathbf{b}$ have a solution $\mathbf{x} \geq \mathbf{0}$, \qquad (13)

or

the inequalities $\mathbf{y}'A \leq \mathbf{0}$, $\mathbf{y}'\mathbf{b} > 0$ have a solution $\mathbf{y} \geq \mathbf{0}$, \qquad (14)

but not both.

*Proof.* If both (13) and (14) hold, then

$$0 \geq \mathbf{y}'A\mathbf{x} \geq \mathbf{y}'\mathbf{b} > 0 ,$$

which is impossible. If (13) is false, then

$A\mathbf{x} - \mathbf{z} = \mathbf{b}$ has no solution for which $\mathbf{x} \geq \mathbf{0}$ and $\mathbf{z} \geq \mathbf{0}$,

that is $(A, -I) \begin{pmatrix} \mathbf{x} \\ \mathbf{z} \end{pmatrix} = \mathbf{b}$ has no solution $\begin{pmatrix} \mathbf{x} \\ \mathbf{z} \end{pmatrix} \geq \mathbf{0}$.

So by Theorem 5, $\mathbf{y}'(A, -I) \leq \mathbf{0}, \mathbf{y}'\mathbf{b} > 0$ have a solution, which is a solution of the inequalities in (14).

The following example uses Theorem 5 and also gives a geometrical interpretation of the theorem.

*Example.* Show that the equations

$$2x_1 + 3x_2 - 5x_3 = -4,$$
$$x_1 + 2x_2 - 6x_3 = -3,$$

have no non-negative solution.

*Solution.* If we can exhibit a solution of the inequalities

$$2y_1 + y_2 \leq 0, \quad 3y_1 + 2y_2 \leq 0, \quad -5y_1 - 6y_2 \leq 0, \quad -4y_1 - 3y_2 > 0,$$

then by Theorem 5, the equations have no non-negative solution. The above inequalities are equivalent to

$$y_1 \leq -\frac{1}{2}y_2, \quad y_1 \leq -\frac{2}{3}y_2, \quad y_1 \geq -\frac{6}{5}y_2, \quad y_1 < -\frac{3}{4}y_2,$$

which are satisfied by $y_1 = -1, y_2 = 1$.

One can see geometrically that the equations have no non-negative solution by writing them in the form

$$x_1\mathbf{a}_1 + x_2\mathbf{a}_2 + x_3\mathbf{a}_3 = \mathbf{b},$$

where $\quad \mathbf{a}_1 = \begin{pmatrix} 2 \\ 1 \end{pmatrix}, \quad \mathbf{a}_2 = \begin{pmatrix} 3 \\ 2 \end{pmatrix}, \quad \mathbf{a}_3 = \begin{pmatrix} -5 \\ -6 \end{pmatrix} \quad$ and $\quad \mathbf{b} = \begin{pmatrix} -4 \\ -3 \end{pmatrix}.$

The set of vectors $\left\{ \sum_{i=1}^{3} x_i\mathbf{a}_i \mid \mathbf{x} \geq \mathbf{0} \right\}$ is represented by the shaded region C in figure 4 and it is seen that $\mathbf{b}$ does not belong to C.

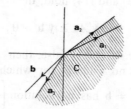

FIGURE 4

# EXERCISES ON CHAPTER ONE

1. Establish whether the following sets are convex or not:

$$\{\mathbf{x} \mid A\mathbf{x} \geq \mathbf{b}, \mathbf{x} \geq \mathbf{0}\}, \quad \{(x,y) \mid xy \leq 1\}, \quad \{\mathbf{x} \mid \mathbf{x}'\mathbf{x} < 1\}.$$

2. Prove that $\langle X \rangle$ is convex and show that $\langle X \rangle$ is the intersection of all the convex sets which contain $X$.

3. Determine the nature of the convex set of the non-negative solutions of the equations, $x_1 + \alpha x_3 = 1$, $x_1 - x_2 = \alpha - 1$, for all values of $\alpha$.

4. If $\mathbf{x} \geq \mathbf{0}$ is a non-basic solution of the equations $A\mathbf{x} = \mathbf{b}$, prove that $\mathbf{x}$ is not a vertex of the convex set, $C = \{\mathbf{x} \mid A\mathbf{x} = \mathbf{b}; \mathbf{x} \geq \mathbf{0}\}$.

5. If $X \subset R^n$ and $\mathbf{x} \in \langle X \rangle$, show that $\mathbf{x}$ can be expressed as a convex combination of not more than $n + 1$ points of $X$. (Use Theorem 3.)

6. If the equations $A\mathbf{x} = \mathbf{0}$ have no non-trivial, non-negative solution, show that any non-negative solution of the equations $A\mathbf{x} = \mathbf{b}$ can be expressed as a convex combination of the basic non-negative solutions of the equations. (Use an inductive proof similar to that of Theorem 3.)

7. Use the theorem of the separating hyperplane to prove that either the equations $A\mathbf{x} = \mathbf{b}$ have a solution, or the equations $\mathbf{y}'A = \mathbf{0}$, $\mathbf{y}'\mathbf{b} = 1$ have a solution, but not both.

8. Show that the inequalities

$$-2x_1 + x_2 - 4x_3 \geq 3, \quad 3x_1 - 2x_2 + 7x_3 \geq -5,$$

have no non-negative solution.

# CHAPTER TWO

# The Theory of Linear Programming

## 1. Examples and Classes of Linear Programmes

The diet problem was one of the problems which led to the development of the theory of Linear Programming. It was first described by Stiegler in a paper published in 1945 on 'The Cost of Subsistence'. In this chapter and in chapter four, we shall use the problem to motivate or interpret some of the definitions, theorems and techniques.

### The Diet Problem

A hospital dietician has to plan a week's diet for a patient, which must include at least a prescribed amount of the nutrients, $N_1, N_2, \ldots, N_m$. He has $n$ different foods, $F_1, F_2, \ldots, F_n$, at his disposal of which he knows the nutritional value and cost. Financial restrictions require him to minimise the cost of the food used in the diet.

If

$a_{ij}$ = the number of units of $N_i$ in one unit of $F_j$,
$b_i$ = the minimum number of units of $N_i$ to be included in the diet,
$c_i$ = the cost of one unit of $F_i$,

then the dietician's problem is to choose $x_i$ units of $F_i$, so that the cost of the diet,

$$c_1x_1 + c_2x_2 + \ldots\ldots + c_nx_n \text{ is a minimum,}$$

14

subject to the diet containing at least $b_i$ units of $N_i$, that is the $x_i$ must satisfy

$$a_{i1}x_1 + a_{i2}x_2 + \ldots\ldots + a_{in}x_n \geq b_i \text{ for } i = 1, 2, \ldots\ldots, m.$$

The remaining constraint,

$$x_1 \geq 0, \; x_2 \geq 0, \ldots, x_n \geq 0,$$

says that the amounts of food in the diet must be non-negative.

The reader should note that the above formulation of the diet problem is a mathematical model and is not altogether realistic, for instance, the cost of food is not necessarily a linear function of the amount required. However, as in other branches of Applied Mathematics, we seek a close approximation to reality, which is tractable.

The diet problem provides an example of the following class of linear programming problems.

*Definition of a standard minimum problem* (s.m.p.)

Minimise $\mathbf{c}'\mathbf{x}$,

subject to $A\mathbf{x} \geq \mathbf{b}$ and $\mathbf{x} \geq \mathbf{0}$,

where $\mathbf{x} = (x_i)$ is a variable *n*-vector, $\mathbf{c} = (c_i)$ is a given *n*-vector, $\mathbf{b} = (b_i)$ is a given *m*-vector and $A = (a_{ij})$ is a given $m \times n$ matrix.

The above notation will be assumed in the subsequent theory, unless the terms are defined alternatively.

Corresponding to any linear programme, there exists a related linear programme, called the dual programme, which is now defined for the s.m.p.

*Definition of the dual of the s.m.p.* (the standard maximum problem)

Maximise $\mathbf{y}'\mathbf{b}$.

subject to $\mathbf{y}'A \leq \mathbf{c}'$ and $\mathbf{y} \geq \mathbf{0}$,

where $\mathbf{y} = (y_i)$ is a variable *m*-vector.

It is easily shown that the dual of the dual is the original s.m.p., by writing the dual in the form

minimise $- \mathbf{b'y}$,
subject to $- A'\mathbf{y} \geq - \mathbf{c}$ and $\mathbf{y} \geq \mathbf{0}$.

The dual programme can often be given a useful economic interpretation, as we now illustrate.

*The dual of the diet problem*

On hearing of the dietician's problem, an enterprising chemist sees his opportunity of making a fortune by manufacturing nutrient pills. He considers that the dietician will be prepared to substitute pills for food in his diet, if the pills are no more expensive than the food. The chemist's problem is to choose the price $y_i$ of one unit of the $N_i$ pill, so that his receipts, based on the minimal nutritional requirements of the diet, are maximised. This is equivalent to

$$\text{maximising } y_1 b_1 + y_2 b_2 + \ldots + y_m b_m.$$

The constraints on the $y_i$ are

$$y_1 a_{1j} + y_2 a_{2j} + \ldots + y_m a_{mj} \leq c_j \quad \text{for } i = 1, 2, \ldots, n,$$

which say that the equivalent pill form of each food is not more expensive than the food itself, and

$$y_1 \geq 0, y_2 \geq 0, \ldots, y_m \geq 0,$$

which is the condition that the prices be non-negative.

The maximising value of $y_i$ can be used as the imputed cost of one unit of the $i$th nutrient, which is the maximum price that the dietician would be prepared to pay for one unit of the $i$th nutrient.

Another useful class of linear programming problems are the canonical problems, as all problems must be put into canonical form to apply the simplex method of solution.

*Definition of a canonical minimum problem* (c.m.p.)

Minimise $\mathbf{c'x}$,
subject to $A\mathbf{x} = \mathbf{b}$ and $\mathbf{x} \geq \mathbf{0}$.

To determine the dual problem, which is consistent with the definition of the dual of the s.m.p., we convert the c.m.p. into the following s.m.p.,

minimise $\mathbf{c'x}$,

subject to $\begin{pmatrix} A \\ -A \end{pmatrix} \mathbf{x} \leq \begin{pmatrix} \mathbf{b} \\ -\mathbf{b} \end{pmatrix}$ and $\mathbf{x} \geq \mathbf{0}$.

The corresponding dual problem is to

maximise $\mathbf{z'b} - \mathbf{w'b}$,
subject to $\mathbf{z'}A - \mathbf{w'}A \leq \mathbf{c'}$ and $\mathbf{z} \geq \mathbf{0}$, $\mathbf{w} \geq \mathbf{0}$.

where $\mathbf{z}$ and $\mathbf{w}$ are $m$-vectors. Since any vector can be expressed as the difference between two non-negative vectors, we write $\mathbf{y} = \mathbf{z} - \mathbf{w}$ to obtain the dual of the c.m.p.

*Dual of the canonical minimum problem*

Maximise $\mathbf{y'b}$,
subject to $\mathbf{y'}A \leq \mathbf{c'}$.

The reader should note that the vector $\mathbf{y}$ in the dual problem is unrestricted in sign. We now show that any linear programming problem can be expressed as a c.m.p. Consider the general linear programming problem:

maximise $c_1x_1 + c_2x_2 + \ldots + c_nx_n$,

subject to 
$$\sum_{j=1}^{n} a_{ij}x_j \leq b_i \qquad \text{for } i = 1, \ldots, r, \qquad (1)$$

$$\sum_{j=1}^{n} a_{ij}x_j \geq b_i \qquad \text{for } i = r + 1, \ldots, s, \qquad (2)$$

$$\sum_{j=1}^{n} a_{ij}x_j = b_i \qquad \text{for } i = s + 1, \ldots, m$$

17

and $x_i \geq 0$ for $i \in S$, where $S$ is a subset of the integers $\{1, 2, \ldots, n\}$. The inequalities (1) and (2) are equivalent to the equations

$$\sum_{j=1}^{n} a_{ij}x_j + w_i = b_i \quad \text{for } i = 1, \ldots, r,$$

$$\sum_{j=1}^{n} a_{ij}x_j - w_i = b_i \quad \text{for } i = r + 1, \ldots, s,$$

where $w_i \geq 0$ for $i = 1, 2, \ldots, s$. In order that all the variables are non-negative, we put $x_i = u_i - v_i$ for $i \notin S$, where $u_i \geq 0$ and $v_i \geq 0$. Combining the above steps gives the following c.m.p.:

minimise $\quad -\sum_{i \in S} c_i x_i - \sum_{i \notin S} c_i(u_i - v_i)$,

subject to $\quad \sum_{j \in S} a_{ij}x_j + \sum_{j \notin S} a_{ij}(u_j - v_j) + w_i = b_i$ for $i = 1, \ldots, r$,

$\quad \sum_{j \in S} a_{ij}x_j + \sum_{j \notin S} a_{ij}(u_j - v_j) - w_i = b_i$ for $i = r + 1, \ldots, s$,

$\quad \sum_{j \in S} a_{ij}x_j + \sum_{j \notin S} a_{ij}(u_j - v_j) = b_i \quad$ for $i = s + 1, \ldots, m$

and $x_i \geq 0$ for $i \in S$, $u_i \geq 0$ for $i \notin S$, $v_i \geq 0$ for $i \notin S$ and $w_i \geq 0$ for $i = 1, 2, \ldots, s$. The derivation of the dual problem is left as an exercise to the reader.

## 2. Fundamental Duality Theorem

DEFINITIONS. A vector satisfying the constraints of a linear programme is called a feasible vector. A feasible vector which maximises (minimises) the given linear function, known as the objective function, is called an optimal vector and the value of the maximum (minimum) is called the value of the programme.

LEMMA 1. If $\mathbf{x}$ and $\mathbf{y}$ are feasible vectors for the s.m.p. and its dual, respectively, then

$$\mathbf{y}'\mathbf{b} \leq \mathbf{y}'A\mathbf{x} \leq \mathbf{c}'\mathbf{x}.$$

*Proof.* Since $\mathbf{x}$ and $\mathbf{y}$ are feasible, $\mathbf{y} \geq \mathbf{0}$ and $A\mathbf{x} \geq \mathbf{b}$, which together imply $\mathbf{y}'A\mathbf{x} \geq \mathbf{y}'\mathbf{b}$.

Also $\mathbf{x} \geq \mathbf{0}$ and $\mathbf{y}'A \leq \mathbf{c}' \implies \mathbf{y}'A\mathbf{x} \leq \mathbf{c}'\mathbf{x}$.

In terms of the diet problem, the above result says that the cost of the diet using pills does not exceed the cost of the diet using food.

THEOREM 2. If $\mathbf{x}^*$ and $\mathbf{y}^*$ are feasible vectors for the s.m.p. and its dual, respectively, which also satisfy

$$\mathbf{c}'\mathbf{x}^* = \mathbf{y}^{*\prime}\mathbf{b},$$

then $\mathbf{x}^*$ and $\mathbf{y}^*$ are optimal vectors.

*Proof.* Let $\mathbf{x}$ be any other feasible vector for the s.m.p., then by Lemma 1, $\mathbf{c}'\mathbf{x} \geq \mathbf{y}^{*\prime}\mathbf{b} = \mathbf{c}'\mathbf{x}^*$. Hence $\mathbf{x}^*$ minimises $\mathbf{c}'\mathbf{x}$ and is an optimal vector. The proof for $\mathbf{y}^*$ follows similarly.

The above theorem gives a sufficient condition for optimality and holds for any linear programme and its dual. The following theorem, which is the basic theorem of linear programming, proves that a necessary and sufficient condition for the existence of optimal vectors is that a linear programme and its dual should be feasible. It also proves that the condition in Theorem 2 is necessary for optimality.

THEOREM 3 (fundamental duality theorem). If both a linear programme and its dual are feasible, then both have optimal vectors and the values of the two programmes are equal.

If either programme has no feasible vector, then neither has an optimal vector.

*Proof.* Since all linear programmes can be expressed as a s.m.p., there is no loss of generality in proving the theorem for the s.m.p. The first part of the theorem will be established if we can show that there exist vectors $\mathbf{x}^* \geq \mathbf{0}$ and $\mathbf{y}^* \geq \mathbf{0}$ satisfying

$$A\mathbf{x} \geq \mathbf{b} \tag{3}$$

$$\mathbf{y}'A \leq \mathbf{c}' \tag{4}$$

$$\mathbf{c}'\mathbf{x} \leq \mathbf{y}'\mathbf{b}; \tag{5}$$

for then by Lemma 1, $\mathbf{c}'\mathbf{x}^* = \mathbf{y}^{*\prime}\mathbf{b}$ and the values of the two programmes are equal. The optimality of $\mathbf{x}^*$ and $\mathbf{y}^*$ immediately follows from Theorem 2.

The system of inequalities (3), (4) and (5) can be written as

$$\begin{pmatrix} A & O \\ O & -A' \\ -\mathbf{c}' & \mathbf{b}' \end{pmatrix} \begin{pmatrix} \mathbf{x} \\ \mathbf{y} \end{pmatrix} \geq \begin{pmatrix} \mathbf{b} \\ -\mathbf{c} \\ 0 \end{pmatrix}. \tag{6}$$

Suppose (6) has no solution for which $\mathbf{x} \geq \mathbf{0}$ and $\mathbf{y} \geq \mathbf{0}$, then by Theorem 6 on page 11 there exist vectors $\mathbf{z} \geq \mathbf{0}$, $\mathbf{w} \geq \mathbf{0}$ and a scalar $\lambda \geq 0$ satisfying the inequalities

$$(\mathbf{z}',\mathbf{w}',\lambda) \begin{pmatrix} A & O \\ O & -A' \\ -\mathbf{c}' & \mathbf{b}' \end{pmatrix} \leq 0, \quad (\mathbf{z}',\mathbf{w}',\lambda) \begin{pmatrix} \mathbf{b} \\ -\mathbf{c} \\ 0 \end{pmatrix} > 0,$$

which can be written as

$$\mathbf{z}'A \leq \lambda\mathbf{c}' \tag{7}$$

$$A\mathbf{w} \geq \lambda\mathbf{b} \tag{8}$$

$$\mathbf{z}'\mathbf{b} > \mathbf{c}'\mathbf{w}. \tag{9}$$

If $\lambda = 0$, then since the s.m.p. and its dual are feasible, there exist vectors $\mathbf{x} \geq \mathbf{0}$ and $\mathbf{y} \geq \mathbf{0}$ satisfying (3) and (4) respectively. It now follows from (3) and (7) that

$$0 \geq \mathbf{z}'A\mathbf{x} \geq \mathbf{z}'\mathbf{b}$$

20

and from (4) and (8) that

$$0 \leq \mathbf{y}'A\mathbf{w} \leq \mathbf{c}'\mathbf{w},$$

which together imply that $\mathbf{z}'\mathbf{b} \leq \mathbf{c}'\mathbf{w}$, contradicting (9). Hence $\lambda > 0$ and $\lambda^{-1}\mathbf{w}$ and $\lambda^{-1}\mathbf{z}$ are feasible vectors for the s.m.p. and its dual, respectively, so by Lemma 1, $\lambda^{-1}\mathbf{z}'\mathbf{b} \leq \lambda^{-1}\mathbf{c}'\mathbf{w}$, which again contradicts (9). Therefore our original assumption is false and the inequalities (6) have a non-negative solution, which provides the required optimal vectors.

To prove the second part of the theorem, we suppose that the s.m.p. has no feasible vector, that is $A\mathbf{x} \geq \mathbf{b}$ has no solution for which $\mathbf{x} \geq \mathbf{0}$, and so by Theorem 6 on page 11, the inequalities $\mathbf{y}'A \leq \mathbf{0}$, $\mathbf{y}'\mathbf{b} > 0$ have a solution $\mathbf{y}_1 \geq \mathbf{0}$. If the dual has no feasible vector, there is nothing to prove, so suppose $\mathbf{y}_0$ is a feasible vector for the dual, then it satisfies

$$\mathbf{y}'A \leq \mathbf{c}', \mathbf{y}' \geq \mathbf{0}. \tag{10}$$

The vector $\mathbf{y}_0 + \lambda \mathbf{y}_1$ also satisfies (10) for all $\lambda \geq 0$ and so is a feasible vector. Since $\mathbf{y}_1'\mathbf{b} > 0$, $(\mathbf{y}_0 + \lambda \mathbf{y}_1)'\mathbf{b}$ can be made arbitrarily large by choosing $\lambda$ sufficiently large, so $\mathbf{y}'\mathbf{b}$ is unbounded above on the set of feasible vectors and the dual has no optimal vector. It follows from the dual relationship that if the dual has no feasible vector, then the s.m.p. has no optimal vector and the proof is completed.

Since the optimal vectors, $\mathbf{x}^*$ and $\mathbf{y}^*$ of the programme and its dual, respectively, satisfy $\mathbf{c}'\mathbf{x}^* = \mathbf{y}^{*\prime}\mathbf{b}$, the value of $y_i^*$ can be used to determine how critical the value of the programme is to small changes in $b_i$. For example, if $y_1^*$ is large compared with the remaining $y_i^*$'s, then a small change in $b_1$ could make a large difference to the minimum value of $\mathbf{c}'\mathbf{x}$.

We now deduce two theorems, which can be used to test given feasible vectors for optimality.

## 3. Equilibrium Theorems

THEOREM 4 (standard equilibrium theorem). The feasible vectors $\mathbf{x}$ and $\mathbf{y}$ of the s.m.p. and its dual, respectively, are optimal if and only if

$$y_i > 0 \Rightarrow (A\mathbf{x})_i = b_i \tag{11}$$

and
$$x_i > 0 \Rightarrow (\mathbf{y}'A)_i = c_i. \tag{12}$$

The conditions (11) and (12) are respectively equivalent to

$$(A\mathbf{x})_i > b_i \Rightarrow y_i = 0 \tag{13}$$

and
$$(\mathbf{y}'A)_i < c_i \Rightarrow x_i = 0. \tag{14}$$

*Proof.* Suppose $\mathbf{x}$ and $\mathbf{y}$ are feasible vectors which satisfy (11) and (12), then

$$\mathbf{y}'A\mathbf{x} = \sum_{i=1}^{m} y_i(A\mathbf{x})_i = \sum_{i=1}^{m} y_i b_i = \mathbf{y}'\mathbf{b}$$

and
$$\mathbf{y}'A\mathbf{x} = \sum_{i=1}^{n} (\mathbf{y}'A)_i x_i = \sum_{i=1}^{n} c_i x_i = \mathbf{c}'\mathbf{x}.$$

Hence by Theorem 2, $\mathbf{x}$ and $\mathbf{y}$ are optimal.

Conversely suppose $\mathbf{x}$ and $\mathbf{y}$ are optimal, then by Theorem 3 and Lemma 1

$$\mathbf{y}'\mathbf{b} = \mathbf{y}'A\mathbf{x} = \mathbf{c}'\mathbf{x}. \tag{15}$$

On writing the left-hand side of (15) in the form

$$\sum_{i=1}^{m} y_i((A\mathbf{x})_i - b_i) = 0$$

22

and using the inequalities $\mathbf{y} \geq \mathbf{0}$ and $A\mathbf{x} \geq \mathbf{b}$, we deduce that

$$y_i((A\mathbf{x})_i - b_i) = 0 \text{ for all } i,$$

which gives the conditions (11) and (13). The conditions (12) and (14) follow on writing the right-hand side of (15) as

$$\sum_{i=1}^{n} (c_i - (\mathbf{y}'A)_i)x_i = 0.$$

In terms of the diet problem, (13) says that if the nutrient $N_i$ is oversupplied in the optimal diet, then the price of one unit of the $N_i$ pill is zero, and (14) says that the amount of the food $F_i$ in the optimal diet is zero, if it is more expensive than its pill equivalent. This suggests that the food $F_i$ is over-priced.

THEOREM 5 (canonical equilibrium theorem). The feasible vectors $\mathbf{x}$ and $\mathbf{y}$ of the c.m.p. and its dual, respectively, are optimal if and only if

$$x_i > 0 \Rightarrow (\mathbf{y}'A)_i = c_i \tag{16}$$

or equivalently $(\mathbf{y}'A)_i < c_i \Rightarrow x_i = 0.$ (17)

*Proof.* Suppose $\mathbf{x}$ and $\mathbf{y}$ are feasible vectors which satisfy (16), then

$$\mathbf{y}'A\mathbf{x} = \sum_{i=1}^{n} (\mathbf{y}'A)_i x_i = \sum_{i=1}^{n} c_i x_i = \mathbf{c}'\mathbf{x}.$$

Since $A\mathbf{x} = \mathbf{b}$, $\mathbf{y}'\mathbf{b} = \mathbf{c}'\mathbf{x}$ and the optimality of $\mathbf{x}$ and $\mathbf{y}$ follows from Theorem 2.

Conversely suppose $\mathbf{x}$ and $\mathbf{y}$ are optimal, then by Theorem 3

$$\mathbf{y}'\mathbf{b} = \mathbf{y}'A\mathbf{x} = \mathbf{c}'\mathbf{x}.$$

On writing the right-hand side of the above equation as

$$\sum_{i=1}^{n} (c_i - (\mathbf{y}'A)_i)x_i = 0$$

and using the inequalities $\mathbf{x} \geq \mathbf{0}$ and $\mathbf{y}'A \leq \mathbf{c}'$, we deduce that

$$((\mathbf{y}'A)_i - c_i)x_i = 0 \text{ for all } i,$$

which implies (16).

The following example illustrates how the equilibrium theorems can be used to test a feasible vector for optimality.

*Example.* Verify that $\mathbf{y} = (2,2,4,0)'$ is an optimal vector for the following linear programme.

Maximise $\quad 2y_1 + 4y_2 + y_3 + y_4,$

subject to $\quad y_1 + 3y_2 \quad\quad + y_4 \leq 8,$

$$2y_1 + y_2 \quad\quad\quad\quad \leq 6,$$

$$y_2 + y_3 + y_4 \leq 6,$$

$$y_1 + y_2 + y_3 \quad\quad \leq 9,$$

and $\quad \mathbf{y} \geq \mathbf{0}.$

*Solution.* It is easy to verify that $\mathbf{y} = (2,2,4,0)'$ is feasible for the given standard maximum problem. The dual problem is the following s.m.p.

Minimise $\quad 8x_1 + 6x_2 + 6x_3 + 9x_4,$

subject to $\quad x_1 + 2x_2 \quad\quad + x_4 \geq 2$

$$3x_1 + x_2 + x_3 + x_4 \geq 4$$

$$x_3 + x_4 \geq 1$$

$$x_1 \quad\quad + x_3 \quad\quad \geq 1$$

and $\quad \mathbf{x} \geq \mathbf{0}.$

24

Assume that $\mathbf{y}$ is optimal, then by Theorem 4, condition (11),

$$y_1 > 0 \Rightarrow \quad x_1 + 2x_2 \qquad\quad + x_4 = 2,$$

$$y_2 > 0 \Rightarrow 3x_1 + \quad x_2 + x_3 + x_4 = 4,$$

$$y_3 > 0 \Rightarrow \qquad\qquad\qquad x_3 + x_4 = 1,$$

and by condition (14),

$$y_1 + y_2 + y_3 < 9 \Rightarrow x_4 = 0.$$

The above equations have the solution $\mathbf{x} = \left(\dfrac{4}{5}, \dfrac{3}{5}, 1, 0\right)' \geq 0$ and since

$x_1 + x_3 = \dfrac{9}{5} \geq 1$, $\mathbf{x}$ is feasible for the dual. Optimality follows since $\mathbf{x}$ and $\mathbf{y}$ are feasible and satisfy the conditions (11) and (14) of Theorem 4.

As a check, we note that

$$\mathbf{c}'\mathbf{x} = 8 \times \frac{4}{5} + 6 \times \frac{3}{5} + 6 \times 1 + 9 \times 0 = 16$$

$$= 2 \times 2 + 4 \times 2 + 1 \times 4 + 1 \times 0 = \mathbf{y}'\mathbf{b}$$

## 4. Basic Optimal Vectors

In Chapter 1 we proved that a linear function on a convex polytope attains its maximum (minimum) at a vertex of the polytope (see Theorem 2 on page 3). We also showed on page 5 that the vertices of the convex set, $C = \{\mathbf{x} \mid A\mathbf{x} = \mathbf{b}, \mathbf{x} \geq 0\}$, are the basic non-negative solutions of the equations, $A\mathbf{x} = \mathbf{b}$. If $C$ is not a convex polytope, then the linear function might not be bounded above (below) on $C$, in which case the maximum (minimum) will not exist. However the next theorem proves that if a linear function attains its maximum (minimum) in $C$, then it attains it at a vertex of $C$.

THEOREM 6. If a canonical minimum problem has an optimal vector, then it has a basic optimal vector. (A basic optimal vector is an optimal vector, which is a basic solution of the constraint equation.)

25

*Proof.* We recall that the constraint equation, $A\mathbf{x} = \mathbf{b}$, can be expressed in the form

$$\sum_{i=1}^{n} x_i \mathbf{a}_i = \mathbf{b}, \tag{18}$$

where $\mathbf{a}_1, \mathbf{a}_2, \ldots, \mathbf{a}_n$ are the column vectors of $A$. Let $\mathbf{x}^*$ be an optimal vector for the c.m.p., then without loss of generality we can assume that

$$x_i^* > 0 \quad \text{for } i = 1, 2, \ldots, r \quad \text{and} \quad x_i^* = 0 \quad \text{for } i = r + 1, \ldots, n,$$

by reordering the $x_i^*$'s if necessary. If $\mathbf{a}_1, \mathbf{a}_2, \ldots, \mathbf{a}_r$ are linearly independent, $\mathbf{x}^*$ is basic and there is nothing to prove. If not, then it follows from Theorem 3 on page 5 that since $\sum_{i=1}^{r} x_i \mathbf{a}_i = \mathbf{b}$ has a non-negative solution $(x_1^*, x_2^*, \ldots, x_r^*)$, it has a basic non-negative solution $(\bar{x}_1, \bar{x}_2, \ldots, \bar{x}_r)$. Hence there exists a vector $\bar{\mathbf{x}} = (\bar{x}_1, \ldots, \bar{x}_r, 0, \ldots, 0)' \geq \mathbf{0}$ satisfying (18) and such that the set of vectors $\{\mathbf{a}_i \mid \bar{x}_i > 0\}$ are linearly independent, so $\bar{\mathbf{x}}$ is a basic non-negative solution of (18). Let $\mathbf{y}$ be an optimal vector for the dual, then by Theorem 5, since $\mathbf{x}^*$ is optimal

$$(\mathbf{y}'A)_i < c_i \Longrightarrow x_i^* = 0$$

but

$$x_i^* = 0 \Longrightarrow \bar{x}_i = 0,$$

hence $\bar{\mathbf{x}}$ satisfies the condition (17) for optimality and is therefore, a basic optimal vector.

The above theorem gives us a method of finding an optimal vector, which is to solve the system of equations $\sum_{i \in S} x_i \mathbf{a}_i = \mathbf{b}$ for all sets $S$, such that the set of vectors $\{\mathbf{a}_i \mid i \in S\}$ are linearly independent. The non-negative solution which minimises $\mathbf{c}'\mathbf{x}$

is a basic optimal vector, provided an optimal vector exists. However for an $m \times n$ matrix $A$, the number of trials required could reach $\binom{n}{m}$ , which is prohibitive for large $n$ and $m$. The simplex method, described in Chapter 4, is much more efficient.

## 5. Graphical Method of Solution

We conclude this chapter by illustrating how standard linear programming problems in two variables can be solved graphically.

*Example.*

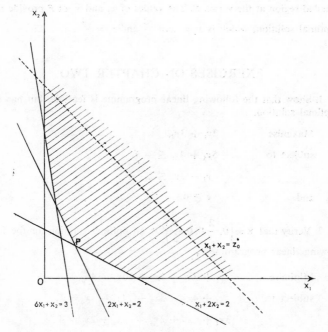

FIGURE 5

Minimise $\quad\quad\quad x_1 + x_2$

subject to $\quad\quad 2x_1 + \phantom{2}x_2 \geq 2$

$$x_1 + 2x_2 \geq 2$$

$$6x_1 + \phantom{2}x_2 \geq 3$$

and $\quad\quad\quad x_1 \geq 0, x_2 \geq 0.$

*Solution.* The values of $x_1$ and $x_2$ satisfying one of the above constraints, lie in the half-plane bounded by the straight line given by satisfying the inequality as an equation. In figure 5, the shaded region represents the intersection of all the halfplanes and is the region of feasible values of $x_1$ and $x_2$. The equation $z_0 = x_1 + x_2$ is a straight line with perpendicular distance $z_0/\sqrt{2}$ from the origin, so to minimise $z_0$, the line is moved parallel to itself towards the origin until it only meets the shaded region at the vertex $P$. The values of $x_1$ and $x_2$ at $P$ provide the optimal solution, which is $x_1 = x_2 = \dfrac{2}{3}$ and $z_0 = \dfrac{4}{3}$.

# EXERCISES ON CHAPTER TWO

1. Show that the following linear programme is feasible, but has no optimal solution.

Maximise $\quad\quad 2y_1 + 3y_2,$

subject to $\quad -5y_1 + 4y_2 \leq -1,$

$$y_1 - \phantom{5}y_2 \leq 2,$$

and $\quad\quad\quad y \geq 0.$

2. Verify that $\mathbf{x} = \left(0, \dfrac{2}{3}, 0, 0, \dfrac{5}{3}\right)'$ is an optimal solution for the following linear programme.

Minimise $\quad\quad x_1 + 7x_2 - 5x_3 + 10x_4 - 4x_5,$

subject to $\quad\quad x_1 + 4x_2 - 3x_3 + \phantom{1}4x_4 - \phantom{2}x_5 = 1,$

$$3x_1 + \phantom{4}x_2 + 2x_3 - \phantom{1}x_4 + 2x_5 = 4.$$

and $\quad\quad\quad \mathbf{x} \geq \mathbf{0}.$

3. The following numbers of waitresses are required each day at a transport cafe during the given periods.

| 3.01 – 7.00 | 7.01 – 11.00 | 11.01 – 3.00 | 3.01 – 7.00 | 7.01 – 11.00 | 11.01 – 3.0 |
|---|---|---|---|---|---|
| 2 | 10 | 14 | 8 | 10 | 3 |

The waitresses report for duty at 3a.m., 7a.m., etc., and their tours of duty last for eight hours. The problem is to determine how the given numbers can be supplied for each period, whilst involving the smallest number of waitresses each day. If $x_1, x_2, \ldots, x_6$ are the numbers starting at 3a.m., 7a.m., ..., 11p.m., respectively, verify that an optimal solution is $x = (0,14,0,8,2,2)'$.

4. Show that the following linear programme has an optimal solution, and find it by computing the basic feasible solutions.

Minimise $\quad 2x_1 + x_2 - 4x_3$,

subject to $\quad x_1 + 3x_2 - x_3 = 1$,

$\qquad\qquad 3x_1 + 2x_2 + x_3 = 7$,

and $\quad x \geq 0$.

5. Show that the dual of the general linear programme, defined on page 18, is to

minimise $\displaystyle\sum_{i=1}^{m} y_i b_i$

subject to $\displaystyle\sum_{i=1}^{m} y_i a_{ij} \geq c_j$ for $j \epsilon S$

$\displaystyle\sum_{i=1}^{m} y_i a_{ij} = c_j$ for $j \notin S$

and $y_i \geq 0$ for $i = 1, \ldots, r$; $y_i \leq 0$ for $i = r + 1, \ldots, s$.

6. Formulate the following problem from Approximation Theory as a linear programme. For given functions $\phi_1(x)$, $\phi_2(x)$, ..., $\phi_n(x)$, find $\alpha_1, \alpha_2, \ldots, \alpha_n$ such that $\displaystyle\sum_{i=1}^{n} \alpha_i \phi_i(x)$ is the best approximation to a function $f(x)$ on the set $\{x_1, x_2, \ldots, x_m\}$ in the sense that the $\alpha_i$ minimise the

29

$\text{maximum} \underset{j}{\mid} f(x_j) - \sum_{i=1}^{n} \alpha_i \phi_i(x_j) \mid$. Show that the dual is a canonical problem.

(Hint: let $\alpha_{n+1} = \max_j \mid f(x_j) - \sum_{i=1}^{n} \alpha_i \phi_i(x_j) \mid$ and minimise $\alpha_{n+1}$.)

7. Give the dual of the following linear programme:

minimise $\qquad \lambda$

subject to $\qquad A\mathbf{x} + \lambda \mathbf{b} \geq \mathbf{b}$ and $\mathbf{x} \geq 0, \ \lambda \geq 0$.

Show that the problem and its dual are feasible, and hence obtain a proof of Theorem 6 on page 11.

8. Wall paper is supplied in rolls of 33 feet length, and 19 strips of 7 feet length and 8 strips of 3 feet length are required to paper a room. Verify that to minimise the number of rolls used, 4 rolls must be cut in the pattern $4 \times 7' + 1 \times 3'$ with $2'$ wasted and 1 roll in the pattern $3 \times 7' + 4 \times 3'$.

9. If $\mathbf{x} = (x_1, x_2, \ldots, x_m, 0, \ldots, 0)'$ is a basic feasible vector for the canonical minimum problem, where $x_i > 0$ for $i = 1, 2, \ldots, m$, and if the vector $\mathbf{y}$, which satisfies the equations $\mathbf{y}'\mathbf{a}_i = c_i$ for $i = 1, 2, \ldots, m$, is not feasible for the dual problem, prove that $\mathbf{x}$ is not optimal.

10. A firm can manufacture $n$ different products in a given period of time. Constraints upon production are imposed by a fixed capacity in each of the $m$ production departments and by minimum and maximum production levels obtained from contracts and sales forecasts, respectively. Let $c_j$ be the profit on one unit of the $j$th product, $b_i$ be the capacity of the $i$th department, $a_{ij}$ be the number of units of capacity of the $i$th department required to produce one unit of the $j$th product and let $u_j$ and $l_j$ be the upper and lower limits, respectively, upon the production of the $j$th product. If $x_j$ units of the $j$th product are manufactured, formulate the problem of maximising the profits as a linear programme.

By considering the dual programme, show that the feasible vector $\mathbf{x} = (x_j)$ is optimal if there exists a vector $\mathbf{y} = (y_i)$ satisfying $(\mathbf{y}'A)_j = c_j$ for $l_j < x_j < u_j$, $(\mathbf{y}'A)_j \leq c_j$ for $x_j = u_j$ and $(\mathbf{y}'A)_j \geq c_j$ for $x_j = l_j$, where $A = (a_{ij})$.

# CHAPTER THREE

# The Transportation Problem

## 1. Formulation of Problem and Dual

The transportation problem was first described by Hitchock in a paper on 'The Distribution of a Product from Several Sources to Numerous Localities', which was published in 1941. Although the problem can be solved by the simplex method, the special form of its constraint matrix leads to alternative techniques of solution, one of which is described in this chapter[1].

### The Transportation Problem

A commodity is manufactured at $m$ factories, $F_1$, $F_2$, ..., $F_m$, and is sold at $n$ markets, $M_1$, $M_2$, ..., $M_n$. The annual output at $F_i$ is $s_i$ units and the annual demand at $M_j$ is $d_j$ units. If the cost of transporting one unit of the commodity from $F_i$ to $M_j$ is $c_{ij}$, then the problem is to determine which factories should supply which markets in order to minimise the transportation costs. For a realistic problem, we can assume $c_{ij} \geq 0$, $s_i > 0$ and $d_j > 0$. If $x_{ij}$ is the number of units transported per year from $F_i$ to $M_j$, then the transportation costs,

$$\sum_{i=1}^{m} \sum_{j=1}^{n} c_{ij} x_{ij} \quad \text{are to be minimised,}$$

[1] Another technique is based on network theory, see Chapter 5 of Gale.

subject to the amount taken from $F_i$ not exceeding the supply there,

$$\text{that is } \sum_{j=1}^{n} x_{ij} \leq s_i \quad \text{for } i = 1, 2, \ldots, m, \tag{1}$$

and the amount taken to $M_j$ satisfying the demand at $M_j$,

$$\text{that is } \sum_{i=1}^{m} x_{ij} \geq d_j \quad \text{for } j = 1, 2, \ldots, n. \tag{2}$$

The amounts carried from $F_i$ to $M_j$ must be non-negative, so

$$x_{ij} \geq 0 \quad \text{for } i = 1, 2, \ldots, m \quad \text{and} \quad j = 1, 2, \ldots, n.$$

It follows from (1) and (2) that for the problem to be feasible,

$$\sum_{i=1}^{m} s_i \geq \sum_{i=1}^{m} \sum_{j=1}^{n} x_{ij} \geq \sum_{j=1}^{n} d_j,$$

so the total demand must not exceed the total supply. If the total supply and demand are equal, then (1) and (2) must be satisfied as equations. In any case, the problem can always be formulated so that the total supply and demand are equal. This is achieved by introducing an extra market $M_{n+1}$ (a dump) with demand $d_{n+1} = \sum_{i=1}^{m} s_i - \sum_{j=1}^{n} d_j$, the excess supply, and by choosing $c_{in+1} = 0$ for $i = 1, \ldots, m$. It can be shown that an optimal solution for the modified problem is an optimal solution for the original problem.

We therefore define the transportation problem as the following c.m.p.

Minimise $\sum\limits_{i=1}^{m} \sum\limits_{j=1}^{n} c_{ij}x_{ij}$,

subject to $\quad \sum\limits_{j=1}^{n} x_{ij} = s_i \qquad$ for $i = 1,2,\ldots,m$, $\qquad$ (3)

$$\sum_{i=1}^{m} x_{ij} = d_j \qquad \text{for } j = 1,2,\ldots,n, \qquad (4)$$

and $\quad x_{ij} \geq 0$ for $i = 1,2,\ldots,m$ and $j = 1,2,\ldots,n$, $\qquad$ (5)

where $s_i > 0$, $d_j > 0$ and $c_{ij} \geq 0$ are given and $\sum\limits_{i=1}^{m} s_i = \sum\limits_{j=1}^{n} d_j$.

*The dual of the transportation problem*

Let $\quad \mathbf{x} = (x_{11}, x_{12}, \ldots, x_{1n}, \ldots, x_{m1}, x_{m2}, \ldots, x_{mn})'$,

$\quad\quad\ \ \mathbf{c} = (c_{11}, c_{12}, \ldots, c_{1n}, \ldots, c_{m1}, c_{m2}, \ldots, c_{mn})'$,

$\quad\quad\ \ \mathbf{b} = (s_1, s_2, \ldots, \ s_m, d_1, d_2, \ldots, d_n)'$,

and the $(m + n) \times mn$ matrix

$$A = (\mathbf{a}_{11}, \mathbf{a}_{12}, \ldots, \mathbf{a}_{1n}, \ldots, \mathbf{a}_{m1}, \mathbf{a}_{m2}, \ldots, \mathbf{a}_{mn}), \qquad (6)$$

where $\mathbf{a}_{ij} = \begin{pmatrix} \mathbf{e}_i \\ \mathbf{e}_j \end{pmatrix}$, $\mathbf{e}_i$ is the $i$th unit $m$-vector and $\mathbf{e}_j$ is the $j$th unit $n$-vector. In the case $m = 2$ and $n = 3$,

$$A = \begin{pmatrix} 1 & 1 & 1 & 0 & 0 & 0 \\ 0 & 0 & 0 & 1 & 1 & 1 \\ 1 & 0 & 0 & 1 & 0 & 0 \\ 0 & 1 & 0 & 0 & 1 & 0 \\ 0 & 0 & 1 & 0 & 0 & 1 \end{pmatrix}.$$

With the above notation, (3), (4) and (5) reduce to the c.m.p.

minimise $\mathbf{c}' \mathbf{x}$,

subject to $A\mathbf{x} = \mathbf{b}$ and $\mathbf{x} \geq \mathbf{0}$.

The corresponding dual problem is to

maximise $\mathbf{y}'\mathbf{b}$,

subject to $\mathbf{y}'A \leq \mathbf{c}'$,

where $\mathbf{y}$ is an $(m + n)$-vector. If we put

$$\mathbf{y} = (-u_1, -u_2, \ldots, -u_m, v_1, v_2, \ldots, v_n)'$$

then $\mathbf{y}'\mathbf{a}_{ij} = -u_i + v_j$ and $\mathbf{y}'\mathbf{b} = -\sum_{i=1}^{m} u_i s_i + \sum_{j=1}^{n} v_j d_j$.

Hence the dual problem is to find $u_i$ and $v_j$ which

$$\text{maximise } \sum_{j=1}^{n} v_j d_j - \sum_{i=1}^{m} u_i s_i \,, \tag{7}$$

subject to $v_j - u_i \leq c_{ij}$ for $i = 1, 2, \ldots, m$ and $j = 1, 2, \ldots, n$.

$$\tag{8}$$

The dual problem can be interpreted in the following way. A road haulage operator puts in a tender for the transportation of the commodity from the factories to the markets. He offers to pay $u_i$ per unit commodity at the factory $F_i$ and to sell the commodity for $v_j$ per unit at the market $M_j$, so that his charge for carrying unit commodity from $F_i$ to $M_j$ is $v_j - u_i$. With this interpretation (7) represents the amount he will receive for the job, which he will try to maximise, and (8) is the condition that his charges are competitive.

## 2. Theorems Concerning Optimal Solutions

We now show that the transportation problem has an optimal solution, by applying the fundamental duality theorem.

THEOREM 1. The transportation problem has an optimal solution.

*Proof.* By inspection $x_{ij} = s_i d_j \left/ \sum_{i=1}^{m} s_i \right.$ satisfies (3), (4) and (5), and $u_i = v_j = 0$ satisfies (8), as $c_{ij} \geq 0$. Hence the problem and its dual are feasible, so both have an optimal solution by Theorem 3 on page 19.

If the commodity is cars, then the optimal solution will only be meaningful if it specifies an integral number of cars to be transported from $F_i$ to $M_j$. The next two theorems prove that if the supplies and demands are integral, then an optimal solution in integers exists. We first recall the definition of a minor of order $r$, which is the determinant formed from a matrix by omitting all but $r$ rows and $r$ columns of the matrix.[1]

THEOREM 2. Any minor of the matrix $A$, defined by (6), takes one of the values $-1, 0, 1$.

*Proof.* The theorem is proved by induction on the order $r$ of the minor and is trivially true for all minors of order 1. Assume that the theorem holds for all minors of order $r = N - 1$ and consider minors of order $N$. Each column of $A$ contains zeros apart from two unit elements, one of which occurs in the first $m$ rows and the other in the last $n$ rows of $A$. If the minor of order $N$ contains two unit elements in each column, then the sum of its rows taken from the first $m$ rows of $A$ equals the sum of its rows taken from the last $n$ rows of $A$, so the rows are linearly dependent and the value of the minor is zero. If not, then the minor contains one column with at most one non-zero unit element. Expanding the minor by this column and using the inductive hypothesis, it follows that the minor takes one of the values $-1, 0, 1$.

[1] See Cohn page 69.

THEOREM 3. If the supplies $s_i$ and the demands $d_j$ are integral, then the transportation problem has an optimal solution in which the $x_{ij}$'s are integers.

*Proof.* It follows from Theorem 1 and Theorem 6 on page 25 that the transportation problem has a basic optimal solution, so we only need to prove that the basic solutions of $A\mathbf{x} = \mathbf{b}$ are integral.

If $\mathbf{x}$ is a basic solution of $A\mathbf{x} = \mathbf{b}$, then the non-zero elements of $\mathbf{x}$ are the solution of a regular system of equations $M\mathbf{z} = \mathbf{b}_0$, where $M$ is the matrix corresponding to a non-vanishing minor of $A$ and $\mathbf{b}_0$ contains those elements of $\mathbf{b}$ corresponding to the rows of $A$ included in $M$. Since the elements of $\mathbf{b}_0$ are integral and the determinant of $M$ equals $\pm 1$, it follows from Cramer's Rule,[1] that the elements of $\mathbf{z}$ and hence of $\mathbf{x}$ are integral.

## 3. Method of Solution with Modifications for Degeneracy

The following property of the matrix $A$ is important, as it underlies the method of solution of the transportation problem given in this chapter.

THEOREM 4. Rank $A = m+n-1$, where $A$ is defined by (6).

*Proof.* Let the rows of $A$ be the $mn$-vectors $\mathbf{r}_1, \ldots, \mathbf{r}_m, \mathbf{r}_{m+1}, \ldots, \mathbf{r}_{m+n}$. By inspection $\sum\limits_{i=1}^{m} \mathbf{r}_i = \sum\limits_{i=m+1}^{m+n} \mathbf{r}_i$, so the rows of $A$ are linearly dependent. Suppose that

$$\sum_{i=1}^{m+n-1} \lambda_i \mathbf{r}_i = \mathbf{0}. \tag{9}$$

Since the matrix formed by deleting the bottom row of $A$ has only one non-zero element in the $n$-th, $2n$-th, $\ldots$, $mn$-th columns, it follows that by taking the $n$-th, $2n$-th, $\ldots$, $mn$-th

---

[1] See Cohn page 66.

components of (9),

$$\lambda_1 = \lambda_2 = \ldots = \lambda_m = 0.$$

The 1st, 2nd, ..., $(n-1)$-th components of (9) now give

$$\lambda_{m+1} = \lambda_{m+2} = \ldots = \lambda_{m+n-1} = 0.$$

Hence $\mathbf{r}_1, \mathbf{r}_2, \ldots, \mathbf{r}_{m+n-1}$ are linearly independent and rank $A = m+n-1$.

Although rank $A = m+n-1$, the system of equations $A\mathbf{x} = \mathbf{b}$ ((3) and (4)) are consistent, as the total supply equals the total demand. A basic solution of $A\mathbf{x} = \mathbf{b}$ will have at most $m+n-1$ non-zero values of $x_{ij}$ and if the problem is degenerate, which occurs when a partial sum of the $s_i$'s equals a partial sum of the $d_j$'s, then a basic solution may have less than $m+n-1$ non-zero $x_{ij}$'s. The following method of solution assumes that the problem is non-degenerate.

## Method of solution of the transportation problem

The method is based on the equilibrium theorem, Theorem 5 on page 23, which says that the feasible solutions $x_{ij}$, $u_i$ and $v_j$ are optimal if and only if

$$x_{ij} > 0 \Longrightarrow v_j - u_i = c_{ij}.$$

The procedure is to find a basic feasible solution of (3) and (4), which contains $m+n-1$ positive $x_{ij}$ for a non-degenerate problem, and then to solve the $m+n-1$ equations

$$v_j - u_i = c_{ij} \quad \text{for } x_{ij} > 0,$$

which determine the $u_i$ and $v_j$ uniquely, provided a value is assigned to one of them, say $u_1 = 0$. If the $u_i$ and $v_j$ so determined, also satisfy

$$v_j - u_i \leq c_{ij} \quad \text{for } x_{ij} = 0,$$

then they are a feasible solution for the dual problem and the optimality of the $x_{ij}$ follows from the equilibrium theorem. If

$$v_l - u_k > c_{kl} \quad \text{for some } x_{kl} = 0,$$

then the basic feasible solution can be improved by making $x_{kl} > 0$ (see exercise 9 on page 30). It is worth noting that there is no loss of generality in putting $u_1 = 0$, as an arbitrary constant can be added to all the $u_i$ and $v_j$ without altering their feasibility, (8), or the value of the function to be maximised, (7).

We now give two methods for finding a basic feasible solution (b.f.s.). The north-west corner method puts $x_{11} = \min(s_1, d_1)$ and then proceeds along successive rows from left to right with each allocation either using up the supplies at $F_i$ or satisfying the demand at $M_j$. The matrix minimum method finds the minimum element in the cost matrix $(c_{ij})$ and allocates as much as possible along that route. It then repeats the process using successive minima until all the supplies are exhausted and the demands are satisfied. It can be shown that both methods give a b.f.s., though the matrix minimum method usually gives a solution which is closer to the optimal solution, so it is the more efficient method.

*Example.* Find an optimal solution for the transportation problem for which the cost matrix

$$(c_{ij}) = \begin{pmatrix} 5 & 5 & 7 \\ 1 & 2 & 1 \end{pmatrix}$$

the supplies are $s_1 = 3$, $s_2 = 5$
and the demands are $d_1 = 4$, $d_2 = 2$, $d_3 = 2$.

*Solution.* We construct a tableau with the costs $c_{ij}$ in the left, hand side of each column and the initial b.f.s. in the right-hand side. The b.f.s.

has been obtained by the north-west corner method and the $\theta$'s should be neglected for the moment. Putting $u_1 = 0$, the $u_i$ and $v_j$ are calculated from the equations $v_j - u_i = c_{ij}$ for $x_{ij} > 0$, for example $v_1 - 0 = c_{11} = 5$ as $x_{11} > 0$. The circled entries in the

| $d_j$ | | 4 | | 2 | | 2 | |
|---|---|---|---|---|---|---|---|
| $s_i$ | $u_i$ | $v_j$ | 5 | | 6 | | 5 |
| 3 | 0 | 5 | $3-\theta$ | 5 | ⑥ $\theta$ | 7 | ⑤ |
| 5 | 4 | 1 | $1+\theta$ | 2 | $2-\theta$ | 1 | 2 |

right-hand side of the columns are the values of $v_j - u_i$ for $x_{ij} = 0$ which are to be compared with the values of $c_{ij}$ in the left-hand side. Since $v_2 - u_1 = 6 > 5 = c_{12}$, the $v_j$ and $u_i$ are not feasible and for a non-degenerate problem, it can be shown that the $x_{ij}$ are not optimal (see exercise 9 on page 30). Let $x_{12} = \theta$, then to satisfy the constraints $x_{11} = 3 - \theta$, $x_{21} = 1 + \theta$, and $x_{22} = 2 - \theta$. This leads to a change in cost

$$\theta(c_{12} - c_{11} + c_{21} - c_{22}) = \theta[c_{12} - (v_1 - u_1) + (v_1 - u_2) - (v_2 - u_2)] =$$
$$\theta(c_{12} + u_1 - v_2) < 0 \text{ for } \theta > 0.$$

A new b.f.s. is obtained by putting $\theta = 2$, which is the maximum value of $\theta$ consistent with $x_{ij} \geq 0$. It should be noted that $\theta$ is first put into a position for which $v_j - u_i > c_{ij}$ and is then added to or subtracted from positive values of $x_{ij}$. This is necessary to ensure that the new feasible solution is basic and that the transportation costs are reduced. On putting $\theta = 2$ and calculating the $u_i$ and $v_j$ as before, the following tableau is obtained.

| $u_i$ | $v_j$ | 5 | | 5 | | 5 | |
|---|---|---|---|---|---|---|---|
| 0 | | 5 | 1 | 5 | 2 | 7 | ⑤ |
| 4 | | 1 | 3 | 2 | ① | 1 | 2 |

This time the $u_i$ and $v_j$ satisfy $v_j - u_i \leq c_{ij}$ for $x_{ij} = 0$, so an optimal solution is

$$(x_{ij}) = \begin{pmatrix} 1 & 2 & 0 \\ 3 & 0 & 2 \end{pmatrix}$$

and the minimum cost is 20.

It is straightforward to show that if an optimal solution satisfies $v_j - u_i < c_{ij}$ for all $x_{ij} = 0$, then it is unique (see exercise 3 on page 66). So in the above example, the optimal solution is unique. However, if $v_l - u_k = c_{kl}$ for some $x_{kl} = 0$, then another optimal solution can be found, if the problem is non-degenerate.

*Degeneracy*

Degeneracy will be formally defined on page 56 and as mentioned on page 37, occurs for the transportation problem when a partial sum of the $s_i$'s equals a partial sum of the $d_j$'s. Under such circumstances a b.f.s. may be obtained in which fewer than $m+n-1$ of the $x_{ij}$s are positive with the result that there are insufficient equations to determine the $u_i$ and $v_j$. This difficulty is overcome by making the problem non-degenerate as follows. Let

$$\bar{s}_i = s_i + \varepsilon \quad \text{for } i = 1,2,\ldots,m,$$

$$\bar{d}_j = d_j \qquad \text{for } j = 1,2,\ldots,n-1,$$

and $\bar{d}_n = d_n + m\,\varepsilon,$

where $\varepsilon$ is a small quantity chosen so that a partial sum of the $\bar{s}_i$'s is not equal to a partial sum of the $\bar{d}_j$'s, though the total supply still equals the total demand. The problem is now solved with supplies $\bar{s}_i$ and demands $\bar{d}_j$ and the optimal solution for the original problem is found by putting $\varepsilon = 0$ into the optimal solution of the perturbed problem.

*Example.* Solve the transportation problem with cost matrix

$$(c_{ij}) = \begin{pmatrix} 1 & 2 & 4 \\ 1 & 1 & 5 \end{pmatrix}$$

supplies $s_1 = 3$, $s_2 = 5$,
and demands $d_1 = 3$, $d_2 = 3$, $d_3 = 2$.

*Solution.* The problem is degenerate as $s_2 = d_1 + d_3$ and the initial b.f.s.

$$(x_{ij}) = \begin{pmatrix} 3 & 0 & 0 \\ 0 & 3 & 2 \end{pmatrix}$$

contains only three non-zero $x_{ij}$. On making the problem non-degenerate according to the above scheme and constructing an initial b.f.s. by the matrix minimum method, the following tableau is obtained.

| | $d_j$ | 3 | | 3 | | $2+2\varepsilon$ | |
|---|---|---|---|---|---|---|---|
| $s_i$ | $u_i$ \ $v_j$ | 1 | | 0 | | 4 | |
| $3+\varepsilon$ | 0 | 1 | $3-\theta$ | 2 | ⓪ | 4 | $\varepsilon+\theta$ |
| $5+\varepsilon$ | $-1$ | 1 | ② $\theta$ | 1 | 3 | 5 | $2+\varepsilon-\theta$ |

Since $v_1 - u_2 > c_{21}$, we put $x_{21} = \theta$ and find that the maximum value of $\theta = 2+\varepsilon$, which gives the next tableau.

| $u_i$ \ $v_j$ | | 1 | | 1 | | 4 | |
|---|---|---|---|---|---|---|---|
| 0 | | 1 | $1-\varepsilon$ | 2 | ① | 4 | $2+2\varepsilon$ |
| 0 | | 1 | $2+\varepsilon$ | 1 | 3 | 5 | ④ |

The $u_i$ and $v_j$ now satisfy $v_j - u_i \leq c_{ij}$ for $x_{ij} = 0$, so putting $\varepsilon = 0$, we obtain the optimal solution

$$(x_{ij}) = \begin{pmatrix} 1 & 0 & 2 \\ 2 & 3 & 0 \end{pmatrix} \text{ with cost} = 14.$$

## 4. Other Problems of Transportation Type

We conclude this chapter by describing two problems which can be transformed into transportation problems and so can be solved by the method described in this chapter.

## The Caterer's Problem

A caterer has to provide napkins for dinners on $m$ successive days and $n_i$ napkins are required for the dinner on the $i$th day. The napkins can either be bought at $b$ pence each, laundered by the following day at $c$ pence each or laundered in three days at $d$ pence each. If $b > c > d$, the caterer has to determine how many napkins should be bought and how many should be sent to each of the laundry services so as to minimise the cost of providing napkins. To formulate the problem as a transportation problem, let the factories $F_1, F_2, \ldots, F_m$, be the baskets of dirty napkins with supplies $n_1, n_2, \ldots, n_m$, respectively, and $F_{m+1}$ be the shop with a supply $\sum_{i=1}^{m} n_i$ of napkins. For the markets $M_1, M_2, \ldots, M_m$, we take the dinners with demands $n_1, n_2, \ldots, n_m$, respectively, and so that the total supply and demand are equal, we introduce a dump $M_{m+1}$, with a demand of $\sum_{i=1}^{m} n_i$. The cost matrix for the problem is shown below.

| $s_i$ $\quad$ $d_j$ | $n_1$ | $n_2$ | $n_3$ | $n_4$ | $n_5$ | $\ldots$ $n_m$ | $\sum_{i=1}^{m} n_i$ |
|---|---|---|---|---|---|---|---|
| $n_1$ | $\infty$ | $c$ | $c$ | $d$ | $d$ | $\ldots d$ | $0$ |
| $n_2$ | $\infty$ | $\infty$ | $c$ | $c$ | $d$ | $\ldots d$ | $0$ |
| $n_3$ | $\infty$ | $\infty$ | $\infty$ | $c$ | $c$ | $\ldots d$ | $0$ |
| $\vdots$ | $\vdots$ | $\vdots$ | $\vdots$ | $\vdots$ | $\vdots$ | | |
| $n_m$ | $\infty$ | $\infty$ | $\infty$ | $\infty$ | $\infty$ | $\ldots \infty$ | $0$ |
| $\sum_{i=1}^{m} n_i$ | $b$ | $b$ | $b$ | $b$ | $b$ | $\ldots b$ | $0$ |

The cost matrix is constructed by assuming that on the fourth day (say) napkins used from the first day have been laundered by the three day service and napkins used from the second and third days have been laundered by the one day service. The infinite costs represent impossible supplies of napkins.

### The Optimal Assignment Problem

There are $m$ candidates $C_1, C_2, \ldots, C_m$ for $n$ jobs $J_1, J_2, \ldots, J_n$ and an efficiency expert assesses the rating of $C_i$ for $J_j$ as $a_{ij} \geq 0$. The problem is to allocate the candidates to the jobs so as to maximise the sum of the ratings. Hence numbers $x_{ij}$, equal to 0 or 1, are required which

maximise $\displaystyle\sum_{i=1}^{m} \sum_{j=1}^{n} a_{ij} x_{ij}$,

subject to $\displaystyle\sum_{j=1}^{n} x_{ij} \leq 1$ for $i = 1, 2, \ldots, m,$ \hfill (10)

and $\displaystyle\sum_{i=1}^{m} x_{ij} \leq 1$ for $j = 1, 2, \ldots, n.$ \hfill (11)

The inequality (10) says that a candidate is assigned to at most one job and (11) is the condition that at most one candidate is allocated to each job. If $m < n$, then $n - m$ fictitious candidates can be introduced having zero rating for all the jobs and if $n < m$, then $m - n$ fictitious jobs can be included for which all the candidates have zero rating. Hence the problem can be modified so that $m = n$ and every candidate is assigned to a job, as the inequality $a_{ij} \geq 0$ implies that there is no advantage in leaving a job unfilled. With these modifications, the problem is to

maximise $\displaystyle\sum_{i=1}^{n} \sum_{j=1}^{n} a_{ij} x_{ij}$

subject to $\quad \sum_{j=1}^{n} x_{ij} = 1 \quad$ for $i = 1,2,\ldots, n$

$$\sum_{i=1}^{n} x_{ij} = 1 \quad \text{for } j = 1,2,\ldots, n$$

and $\quad x_{ij} \geq 0$ for $i = 1,2,\ldots, n$ and $j = 1,2,\ldots, n$.

It follows from Theorems 1,2 and 3, suitably modified, that the above problem has an optimal solution in which the $x_{ij}$ equal 0 or 1,[1] so there is no loss in demanding that the $x_{ij}$ be non-negative.

The problem can now be solved as a degenerate transportation problem except that the dual variables, $u_i$ and $v_j$ must satisfy

$$v_j - u_i \geq a_{ij} \quad \text{for } x_{ij} = 0,$$

in order that the $x_{ij}$ be optimal, as the objective function is to be maximised. In this problem, the matrix maximum method will, in general, provide the most efficient initial b.f.s.

## EXERCISES ON CHAPTER THREE

1. Solve the transportation problem with cost matrix,

$$(c_{ij}) = \begin{pmatrix} 6 & 2 & 4 \\ 3 & 0 & 2 \end{pmatrix}$$

where the supplies are $s_1 = 6$, $s_2 = 4$,
and the demands are $d_1 = 5$, $d_2 = 2$, $d_3 = 3$.

---

[1] This result could be used to justify a monogamous society.

2. A transportation problem has cost matrix,

$$(c_{ij}) = \begin{pmatrix} 3 & 4 & 6 & 3 \\ 3 & 5 & 7 & \lambda \\ 2 & 6 & 5 & 7 \end{pmatrix}$$

supplies $s_1 = 3$, $s_2 = 5$, $s_3 = 7$ and demands $d_1 = 2$, $d_2 = 5$, $d_3 = 4$, $d_4 = 4$. Find the minimum cost as a function of $\lambda$ for $\lambda \geq 0$.

3. A caterer requires 30 napkins on Monday, 40 on Tuesday, 60 on Wednesday and 40 on Thursday. Napkins cost $15d$, can be laundered by the following day for $6d$ and in two days for $3d$. Determine the most economical plan for the caterer.

4. An efficiency expert has rated four candidates $C_i$ for four jobs $J_j$ according to the table below.

|       | $J_1$ | $J_2$ | $J_3$ | $J_4$ |
|-------|-------|-------|-------|-------|
| $C_1$ | 8     | 9     | 5     | 6     |
| $C_2$ | 7     | 1     | 2     | 4     |
| $C_3$ | 6     | 8     | 2     | 5     |
| $C_4$ | 1     | 3     | 9     | 3     |

Find all the optimal assignments of candidates to jobs.

5. Show that the transportation problem with cost matrix $(c_{ij})$ has the same optimal solutions as the transportation problem with cost matrix $(\bar{c}_{ij}) = (c_{ij} + a_i + b_j)$, where $a_1, \ldots, a_m, b_1, \ldots, b_n$ are any numbers. Give an interpretation of this result.

6. A jam manufacturer has to supply $d_1$, $d_2$, $d_3$ and $d_4$ tons of jam, respectively, during the next four months, and his plant can produce up to $s$ tons per month. The cost of producing one ton of jam will be $c_1$, $c_2$, $c_3$ and $c_4$ during these months and there is a storage cost of $k$ per ton per month, if the jam is not delivered in the same month as it is produced. Show that the manufacturer's problem of minimising his costs is e quivalent to a transportation problem and solve the problem when $d_1 = 4$, $d_2 = 7$, $d_3 = 5$, $d_4 = 7$, $s = 6$, $c_1 = 3$, $c_2 = 5$, $c_3 = 4$, $c_4 = 6$ and $k = 1$. (Let $x_{ij}$ be the number of tons produced in the $i$th month for delivery in the $j$th month and let $x_{i5}$ be the unused capacity in the $i$th month.)

7. Show that in any optimal assignment at least one candidate is assigned to the job for which he is best qualified.

# CHAPTER FOUR

# The Simplex Method

## 1. Preliminary Discussion and Rules

The simplex method was invented by Dantzig and was first published in 1951. It can be used to solve any linear programming problem, once it has been put into canonical form. Its name derives from the geometrical 'simplex', as one of the first problems to be solved by the method contained the constraint $\sum_{i=1}^{n+1} x_i = 1$.

Theorem 6 on page 25 says that if a linear programme has an optimal solution, then the optimum is attained at a vertex of the convex set of feasible solutions. Suppose the linear function to be minimised is $\mathbf{c}'\mathbf{x}$, then $z_0 = \mathbf{c}'\mathbf{x}$ is the equation of a hyperplane, orthogonal to $\mathbf{c}$, whose perpendicular distance from the origin is proportional to $z_0$. Hence the optimal solution can be found by drawing hyperplanes, orthogonal to $\mathbf{c}$, through all the vertices and by selecting the vertex whose hyperplane is the least distance from the origin (see figure 5 on page 27). The simplex method chooses a sequence of vertices, such that the value of $z_0$ is reduced at each choice. After a finite number of choices, the minimum, if it exists, is attained.

We begin by recalling the technique of pivotal condensation, which is used at each step of the simplex procedure.

### Pivotal Condensation

Suppose $\mathbf{b}_1, \mathbf{b}_2, \ldots, \mathbf{b}_m$ is a basis for the space $R^m$, then the vectors $\mathbf{a}_1, \mathbf{a}_2, \ldots, \mathbf{a}_n \in R^m$ can be expressed uniquely in terms of the basis by

$$\mathbf{a}_j = \sum_{i=1}^{m} t_{ij}\mathbf{b}_i \quad \text{for } j = 1, 2, \ldots, n. \tag{1}$$

An alternative way of expressing (1), which will be used later, is

$$A = BT, \tag{2}$$

where $A = (\mathbf{a}_1, \mathbf{a}_2, \ldots, \mathbf{a}_n)$, $B = (\mathbf{b}_1, \mathbf{b}_2, \ldots, \mathbf{b}_m)$ and $T = (t_{ij})$. We now represent (1) by the following tableau $T$.

|  | $\mathbf{a}_1$ | $\mathbf{a}_s$ | $\mathbf{a}_n$ |  |  |
|---|---|---|---|---|---|
| $\mathbf{b}_1$ | $t_{11}$ | $t_{1s}$ | $t_{1n}$ | $R_1$ |  |
| $\mathbf{b}_r$ | $t_{r1}$ | $t_{rs}$ | $t_{rn}$ | $R_r$ | $T$ |
| $\mathbf{b}_m$ | $t_{m1}$ | $t_{ms}$ | $t_{mn}$ | $R_m$ |  |

If $t_{rs} \neq 0$, $\mathbf{b}_r$ can be replaced in the basis by $\mathbf{a}_s$ as follows. The relation $\mathbf{a}_s = \sum_{i=1}^{m} t_{is}\mathbf{b}_i$ implies $\mathbf{b}_r = \dfrac{1}{t_{rs}}\mathbf{a}_s - \sum_{\substack{i=1 \\ i \neq r}}^{m} \dfrac{t_{is}}{t_{rs}}\mathbf{b}_i$,

so in terms of the new basis $\mathbf{b}_1, \ldots, \mathbf{b}_{r-1}, \mathbf{a}_s, \mathbf{b}_{r+1}, \ldots, \mathbf{b}_m$,

$$\mathbf{a}_j = \sum_{\substack{i=1 \\ i \neq r}}^{m} \left( t_{ij} - \frac{t_{is}}{t_{rs}} t_{rj} \right)\mathbf{b}_i + \frac{t_{rj}}{t_{rs}}\mathbf{a}_s \quad \text{for } j = 1, 2, \ldots, n,$$

and the new tableau $T^*$ is given below. The element $t_{rs}$ is called

|  | $\mathbf{a}_1$ | $\mathbf{a}_s$ | $\mathbf{a}_n$ |  |  |
|---|---|---|---|---|---|
| $\mathbf{b}_1$ | $t_{11} - \dfrac{t_{1s}t_{r1}}{t_{rs}}$ | $0$ | $t_{1n} - \dfrac{t_{1s}t_{rn}}{t_{rs}}$ | $R_1^* = R_1 - \dfrac{t_{1s}}{t_{rs}} R_r$ |  |
| $\mathbf{a}_s$ | $\dfrac{t_{r1}}{t_{rs}}$ | $1$ | $\dfrac{t_{rn}}{t_{rs}}$ | $R_r^* = \dfrac{1}{t_{rs}} R_r$ | $T^*$ |
| $\mathbf{b}_m$ | $t_{m1} - \dfrac{t_{ms}t_{r1}}{t_{rs}}$ | $0$ | $t_{mn} - \dfrac{t_{ms}t_{rn}}{t_{rs}}$ | $R_m^* = R_m - \dfrac{t_{ms}}{t_{rs}} R_r$ |  |

the pivot and the relationship between the rows $R_i$ of $T$ and the rows $R_i^*$ of $T^*$ is shown in $T^*$. The mechanics of the replacement are summarised by the following rule.

*Replacement Rule*

Using the pivot, $t_{rs} \neq 0$, the tableau $T^*$ is obtained from $T$ by dividing the $r$th row of $T$ by $t_{rs}$ and by subtracting from each other row of $T$, the multiple of the $r$th row which gives a zero in the $s$th column, that is

$$t_{rj}^* = \frac{t_{rj}}{t_{rs}} \quad \text{and} \quad t_{ij}^* = t_{ij} - \frac{t_{is}}{t_{rs}} t_{rj} \quad \text{for} \quad i \neq r.$$

As remarked earlier, the simplex method can only be applied to linear programmes in canonical form, so we shall consider the c.m.p.

minimise $c'x$,

subject to $Ax = b$ and $x \geq 0$,

and following Gale[1], we shall interpret the method in terms of the diet problem.

*An interpretation of the simplex method*

We recall that $x_i$ is the number of units of the food $F_i$ in the diet, $c_i$ is the cost of one unit of $F_i$, $b_i$ is the number of units of the nutrient $N_i$ to be included in the diet and $a_{ij}$ is the number of units of $N_i$ in one unit of $F_j$. Suppose $x = (x_1, \ldots, x_m, 0, \ldots, 0)'$, where $x_i > 0$ for $i = 1, 2, \ldots, m$, is a b.f.s. for the c.m.p., which represents a diet using only the foods $F_1$, $F_2$,

[1] See Gale page 105.

$\ldots, F_m$, then $\sum\limits_{=1}^{m} x_i \mathbf{a}_i = \mathbf{b}$. The remaining columns of $A$ can be expressed uniquely in terms of the basis vectors, $\mathbf{a}_1, \mathbf{a}_2, \ldots, \mathbf{a}_m$, to give the tableau below. We now investigate whether the cost of the diet can be reduced by including the food $F_s$ $(s > m)$ in the diet. From the tableau

$$\mathbf{a}_s = \sum_{i=1}^{m} t_{is}\mathbf{a}_i , \qquad (3)$$

and the right-hand side of (3) can be interpreted as the

| | $\mathbf{c}$ | $c_1$ | $c_m$ | $c_s$ | $c_n$ | |
|---|---|---|---|---|---|---|
| | | $\mathbf{a}_1$ | $\mathbf{a}_m$ | $\mathbf{a}_s$ | $\mathbf{a}_n$ | $\mathbf{b}$ |
| $c_1$ | $\mathbf{a}_1$ | 1 | 0 | $t_{1s}$ | $t_{1n}$ | $x_1$ |
| $c_r$ | $\mathbf{a}_r$ | 0 | 0 | $t_{rs}$ | $t_{rn}$ | $x_r$ |
| $c_m$ | $\mathbf{a}_m$ | 0 | 1 | $t_{ms}$ | $t_{mn}$ | $x_m$ |
| | $\mathbf{z} - \mathbf{c}$ | 0 | 0 | $z_s - c_s$ | $z_n - c_n$ | $z_0$ |

combination of $F_1, F_2, \ldots, F_m$ having the same nutritional content as $F_s$ or in other words substitute $F_s$. The unit cost of substitute $F_s$ is $\sum\limits_{i=1}^{m} t_{is}c_i$ so if $F_s$ is cheaper than its substitute, that is if

$$c_s < \sum_{i=1}^{m} t_{is}c_i = z_s, \qquad \text{(definition of } z_s) \qquad (4)$$

$F_s$ should be included in the diet. Suppose there are $x_s^*$ units of $F_s$ in the new diet, then the amount of $F_i$ is reduced to

$$x_i^* = x_i - t_{is}x_s^* \quad \text{for } i = 1,2,\ldots,m,$$

since it is no longer required to make up substitute $F_s$. For a feasible solution $x_i^* \geq 0$, and the amount of $F_s$ is chosen by

$$x_s^* = \min_{t_{is} > 0} \frac{x_i}{t_{is}} = \frac{x_r}{t_{rs}} \text{ (say)}, \tag{5}$$

so that the new solution is also basic. This means that $F_s$ replaces $F_r$ in the diet.

The simplex method consists of using the criteria (4) and (5) to choose a pivot $t_{rs}$ for the replacement and then applying the replacement rule to obtain a new tableau. The procedure is repeated until no food is cheaper than its substitute, that is until

$$z_j = \sum_i t_{ij} c_i \leq c_j \qquad \text{for } j = 1, 2, \ldots, n. \tag{6}$$

It will be proved on page 53 that when (6) is satisfied, an optimal solution has been obtained. The element $z_0$ in the bottom row, is defined to be the current value of $c'x$, that is

$$z_0 = \sum_{i=1}^m c_i x_i.$$

The following lemma proves that the elements in the bottom row, $z - c$, transform according to the replacement rule under a change of basis.

LEMMA 1. If $t_{rs}$ is the pivot for the replacement, then in the new tableau

$$z_j^* - c_j = z_j - c_j - \frac{(z_s - c_s)}{t_{rs}} t_{rj}$$

and

$$z_0^* = z_0 - \frac{(z_s - c_s)}{t_{rs}} x_r$$

50

*Proof.* By definition

$$z_j^* = \sum_{\substack{i=1 \\ i \neq r}}^{m} t_{ij}^* c_i + t_{rj}^* c_s,$$

and it follows from the replacement rule that

$$z_j^* = \sum_{i=1}^{m} \left( t_{ij} - \frac{t_{is}}{t_{rs}} t_{rj} \right) c_i + \frac{t_{rj}}{t_{rs}} c_s.$$

Hence

$$z_j^* = z_j - \frac{(z_s - c_s)}{t_{rs}} t_{rj},$$

on using the definitions of $z_i$ and $z_s$. The proof for $z_0^*$ follows similarly.

We now formulate the criterion (4) and (5) as two rules for choosing the pivot; the first rule selects the column index and the second, the row index.

*Rule I (for a c.m.p.)*[1]

Calculate the scalar product of the $j$th column of the tableau with the cost vector appropriate to the basis, that is evaluate $z_j = \sum_i t_{ij} c_i$. If $z_j > c_j$ for some $j = s$, include $\mathbf{a}_s$ in the basis.

*Rule II*

Given that $\mathbf{a}_s$ is to be included in the basis, calculate the ratios $\frac{x_i}{t_{is}}$ for $t_{is} > 0$. If the minimum of these ratios is achieved at $i = r$, replace the vector $\mathbf{a}_r$ of the current basis by $\mathbf{a}_s$, using the pivot $t_{rs}$.

[1] If $\mathbf{c}'\mathbf{x}$ is to be maximised, then $z_s < c_s$ is the criterion.

The following simple example, for which there is a b.f.s. by inspection, will help to illustrate the simplex method. In general there is no immediate b.f.s. and a technique for obtaining an initial b.f.s. will be given on page 58.

*Example*

Minimise $\qquad 2x_1 + 6x_2 - 7x_3 + 2x_4 + 4x_5$

subject to $\qquad 4x_1 - 3x_2 + 8x_3 - x_4 \qquad = 12$

$$- x_2 + 12x_3 - 3x_4 + 4x_5 = 20$$

and $\qquad x_i \geq 0.$

*Solution.* By inspection $\mathbf{x} = (3,0,0,0,5)'$ is a b.f.s. and the corresponding basis is $\mathbf{a}_1 = (4,0)'$ and $\mathbf{a}_5 = (0,4)'$. The initial tableau expressing $\mathbf{a}_2$, $\mathbf{a}_3$, $\mathbf{a}_4$ and $\mathbf{b}$ as a linear combination of $\mathbf{a}_1$ and $\mathbf{a}_5$ is shown below.

| | **c** | 2 | 6 | −7 | 2 | 4 | | |
|---|---|---|---|---|---|---|---|---|
| | | $\mathbf{a}_1$ | $\mathbf{a}_2$ | $\mathbf{a}_3$ | $\mathbf{a}_4$ | $\mathbf{a}_5$ | **b** | |
| $c_1 = 2$ | $\mathbf{a}_1$ | 1 | $-\dfrac{3}{4}$ | $\boxed{2}$ | $-\dfrac{1}{4}$ | 0 | 3 | $R_1$ |
| $c_5 = 4$ | $\mathbf{a}_5$ | 0 | $-\dfrac{1}{4}$ | 3 | $-\dfrac{3}{4}$ | 1 | 5 | $R_2$ |
| | $\mathbf{z} - \mathbf{c}$ | 0 | $-\dfrac{17}{2}$ | 23 | $-\dfrac{11}{2}$ | 0 | 26 | $R_3$ |

The bottom row is calculated from the definitions of $z_j - c_j$ and $z_0$, so that for example,

$$z_2 - c_2 = t_{12}c_1 + t_{22}c_5 - c_2 = \left(-\frac{3}{4} \times 2\right) + \left(-\frac{1}{4} \times 4\right) - 6 = -\frac{17}{2}$$

and $\qquad z_0 = x_1 c_1 + x_5 c_5 = (3 \times 2) + (5 \times 4) = 26$

As $z_3 - c_3 = 23 > 0$, $\mathbf{a}_3$ is a candidate for the basis by Rule I, and by Rule II, $t_{13}$ is the pivot for the replacement, as $\dfrac{x_1}{t_{13}} = \dfrac{3}{2} < \dfrac{x_5}{t_{23}} = \dfrac{5}{3}$.

The following tableau is obtained by applying the replacement rule.

|  | $a_1$ | $a_2$ | $a_3$ | $a_4$ | $a_5$ | b |  |
|---|---|---|---|---|---|---|---|
| $a_3$ | $\dfrac{1}{2}$ | $-\dfrac{3}{8}$ | 1 | $-\dfrac{1}{8}$ | 0 | $\dfrac{3}{2}$ | $R_1^* = \dfrac{1}{2} R_1$ |
| $a_5$ | $-\dfrac{3}{2}$ | $\boxed{\dfrac{7}{8}}$ | 0 | $-\dfrac{3}{8}$ | 1 | $\dfrac{1}{2}$ | $R_2^* = R_2 - \dfrac{3}{2} R_1$ |
| $z - c$ | $-\dfrac{23}{2}$ | $\dfrac{1}{8}$ | 0 | $-\dfrac{21}{8}$ | 0 | $-\dfrac{17}{2}$ | $R_3^* = R_3 - \dfrac{23}{2} R_1$ |

This time the pivot is $t_{22}$, as $z_2 - c_2 > 0$ and $t_{22}$ is the only positive $t_{i2}$. On re-applying the replacement rule, we obtain the optimal solution, as $z \le c_j$ for all $j$.

|  | $a_1$ | $a_2$ | $a_3$ | $a_4$ | $a_5$ | b |  |
|---|---|---|---|---|---|---|---|
| $a_3$ |  |  |  |  |  | $\dfrac{12}{7}$ | $R_1^* = R_1 + \dfrac{3}{7} R_2$ |
| $a_2$ |  |  |  |  |  | $\dfrac{4}{7}$ | $R_2^* = \dfrac{8}{7} R_2$ |
| $z - c$ | $-\dfrac{79}{7}$ | 0 | 0 | $-\dfrac{18}{7}$ | $-\dfrac{1}{7}$ | $-\dfrac{60}{7}$ | $R_3^* = R_3 - \dfrac{1}{7} R_2$ |

The optimal vector is $\mathbf{x} = \left(0, \dfrac{4}{7}, \dfrac{12}{7}, 0, 0\right)'$, which corresponds to

the optimal basis $\mathbf{a}_2$ and $\mathbf{a}_3$, and the value of the programme is $z_0 = -\dfrac{60}{7}$.

It is a wise rule to calculate the elements of the bottom row first and then to test for optimality, for in this way, unnecessary computation can be avoided.

## 2. Theory of the Simplex Method

We now prove that (6) is the criterion for optimality.

THEOREM 2 (optimality criterion)[1].

[1] If $\mathbf{c}'\mathbf{x}$ is to be maximised then $z_j \ge c_j$ is the criterion for optimality.

The b.f.s. $\mathbf{x} = (x_1, \ldots, x_m, 0, \ldots, 0)'$ corresponding to the basis $\mathbf{a}_1, \mathbf{a}_2, \ldots, \mathbf{a}_m$, is optimal for the c.m.p. if

$$z_j = \sum_{i=1}^{m} t_{ij} c_i \leq c_j \quad \text{for } j = 1, 2, \ldots, n. \tag{7}$$

where $\mathbf{a}_j = \sum_{i=1}^{m} t_{ij} \mathbf{a}_i \quad \text{for } j = 1, 2, \ldots, n.$ (8)

*Proof.* Since the determinant of $(\mathbf{a}_1, \mathbf{a}_2, \ldots, \mathbf{a}_m)$ is non-zero there exists a unique vector $\mathbf{y}$ satisfying the equations

$$\mathbf{y}' \mathbf{a}_i = c_i \quad \text{for } i = 1, 2, \ldots, m.$$

If (7) is satisfied, then

$$\mathbf{y}' \mathbf{a}_j = \sum_{i=1}^{m} t_{ij} \mathbf{y}' \mathbf{a}_i = \sum_{i=1}^{m} t_{ij} c_i \leq c_j \quad \text{for } j = m + 1, \ldots, n,$$

and $\mathbf{y}$ satisfies the dual constraints, $\mathbf{y}'A \leq \mathbf{c}'$. Since

$$\mathbf{y}' \mathbf{b} = \sum_{i=1}^{n} \mathbf{y}' x_i \mathbf{a}_i = \sum_{i=1}^{m} c_i x_i = \mathbf{c}' \mathbf{x},$$

the optimality of $\mathbf{x}$ and $\mathbf{y}$ follows from Theorem 2 on page 19.

The following alternative proof of Theorem 2 is given as it is independent of the dual problem and also provides further insight into the significance of the $z_j$.

*Alternative Proof*

Let $\mathbf{x}^* = (x_1^*, x_2^*, \ldots, x_n^*)'$ be any feasible vector for the c.m.p., then it follows from (8) that

$$\mathbf{b} = \sum_{i=1}^{n} x_i^* \mathbf{a}_i = \sum_{i=1}^{m} x_i^* \mathbf{a}_i + \sum_{j=m+1}^{n} x_j^* \sum_{i=1}^{m} t_{ij} \mathbf{a}_i.$$

But

$$\mathbf{b} = \sum_{i=1}^{m} x_i \mathbf{a}_i,$$

so equating the coefficients of the linearly independent $\mathbf{a}_1, \mathbf{a}_2, \ldots, \mathbf{a}_m$ gives

$$x_i = x_i^* + \sum_{j=m+1}^{n} t_{ij} x_j^*,$$

and

$$\mathbf{c}'\mathbf{x} = \sum_{i=1}^{m} c_i x_i = \sum_{i=1}^{m} c_i \left( x_i^* + \sum_{j=m+1}^{n} t_{ij} x_j^* \right).$$

Hence it follows from (7) and the non-negativity of $\mathbf{x}^*$, that

$$\mathbf{c}'\mathbf{x} = \mathbf{c}'\mathbf{x}^* + \sum_{j=m+1}^{n} (z_j - c_j) x_j^* \leq \mathbf{c}'\mathbf{x}^*,$$

which proves that $\mathbf{x}$ minimises $\mathbf{c}'\mathbf{x}$.

Some readers will no doubt be asking what happens when Rule II cannot be applied, which is the case when $z_s > c_s$ but none of the $t_{is}$ are positive. The next theorem proves that the problem has no optimal solution if this occurs.

THEOREM 3. If $\mathbf{x} = (x_1, \ldots, x_m, 0, \ldots, 0)'$ is a b.f.s. for the c.m.p. corresponding to the basis $\mathbf{a}_1, \mathbf{a}_2, \ldots, \mathbf{a}_m$, and if for some $s$

$$z_s = \sum_{i=1}^{m} t_{is} > c_s \quad \text{and} \quad t_{is} \leq 0 \quad \text{for} \quad i = 1, 2, \ldots, m,$$

where $\mathbf{a}_s = \sum_{i=1}^{m} t_{is} \mathbf{a}_i$, then the c.m.p. has no optimal solution.

*Proof.* Since $\sum\limits_{i=1}^{m} x_i \mathbf{a}_i = \mathbf{b}$ and $\sum\limits_{i=1}^{m} t_{is} \mathbf{a}_i = \mathbf{a}_s$, it follows that

$$\sum_{i=1}^{m} (x_i - \lambda t_{is}) \mathbf{a}_i + \lambda \mathbf{a}_s = \mathbf{b}$$

and $\mathbf{x}^* = (x_1 - \lambda t_{1s}, \ldots, x_m - \lambda t_{ms}, \; 0, \ldots, \lambda, \ldots, 0)'$ is a feasible vector for all $\lambda \geq 0$, as $t_{is} \leq 0$ for $i = 1, 2, \ldots, m$. On substituting for $\mathbf{x}^*$ we obtain

$$\mathbf{c}'\mathbf{x}^* = \sum_{i=1}^{m} c_i(x_i - \lambda t_{is}) + c_s \lambda = \mathbf{c}'\mathbf{x} - \lambda(z_s - c_s),$$

but $z_s > c_s$ and $\lambda$ can be made arbitrarily large, so $\mathbf{c}'\mathbf{x}$ is unbounded below on the set of feasible vectors. Hence the c.m.p. has no optimal solution.

For the simplex method to be a practical technique for solving linear programmes, it must terminate in a finite number of steps. We now prove that this happens provided the problem is non-degenerate, and a technique for overcoming the difficulty that can arise in degenerate problems, is described on page 63.

*Definition of non-degeneracy*

A canonical linear programme with constraint equation $A\mathbf{x} = \mathbf{b}$, is non-degenerate if $\mathbf{b}$ cannot be expressed as a linear combination of fewer than $m$ columns of $A$, where rank $A = m$.

We can always take the rank of $A$ to equal the number of equations, $m$, for if rank $A = p < m$, then either the system of equations is inconsistent or the number of equations can be reduced to $p$. For example, the last equation of the transportation problem is redundant.

THEOREM 4. If a replacement is made according to Rules I, II and the replacement rule during the solution of a non-degenerate c.m.p., then a new b.f.s. is obtained and the new value of $c'x$ is strictly less than the previous value, that is $z_0^* < z_0$.

*Proof.* Let $t_{rs}$ be the pivot for the replacement and $x = (x_1, \ldots, x_m, 0, \ldots, 0)'$ be the current b.f.s., where $x_i > 0$ for $i = 1, 2, \ldots, m$, as the problem is non-degenerate. It follows from the replacement rule that the new basic solution of $Ax = b$ is

$$x_i^* = x_i - \frac{t_{is}}{t_{rs}} x_r \text{ for } i = 1, 2, \ldots, m, \ x_s^* = \frac{x_r}{t_{rs}} \text{ and } x_i^* = 0$$

otherwise, and feasibility follows from Rule II as

$$\frac{x_r}{t_{rs}} \leq \frac{x_i}{t_{is}} \text{ for } t_{is} > 0 \Rightarrow x_i^* \geq 0 \text{ for all } i.$$

The new value of the programme is given by

$$z_0^* = z_0 - \frac{(z_s - c_s)}{t_{rs}} x_r < z_0$$

as $z_s > c_s$ by Rule I, $t_{rs} > 0$ by Rule II and $x_r > 0$ from the non-degeneracy of the problem.

COROLLARY. For a non-degenerate c.m.p. only a finite number of replacements are required either to obtain an optimal solution or to prove its non-existence.

*Proof.* Since the number of b.f.s.'s is finite and $z_0^* < z_0$ ensures that none of them are repeated, either the criterion of

Theorem 2 or the criterion of Theorem 3 must be reached after a finite number of steps.

In a degenerate problem it is possible for $x_r$ to be zero and then for $z_0^*$ and $z_0$ to be equal, so that in theory one can cycle through a set of distinct bases without reducing $z_0$. However it is usually found in practice that if one replacement fails to reduce $z_0$, then a subsequent one will do so. Geometrically, degeneracy occurs when the $k$-dimensional convex set of feasible vectors has more than $k$ bounding hyperplanes passing through a vertex.

Although the value of a linear programme is unique, the optimal vector is not necessarily unique and it is left as an exercise to the reader to show that if $z_j < c_j$ for all $\mathbf{a}_j$ not in the basis, then the optimal vector is unique (see exercise 3). If $z_s = c_s$ for some $\mathbf{a}_s$ not in the basis and the $\min\limits_{t_{is}>0} \dfrac{x_i}{t_{is}} > 0$, then a replacement can be made which leaves $z_0$ unaltered but gives another optimal vector.

## 3. Further Techniques and Extensions

We now describe a technique for obtaining a b.f.s. and the corresponding simplex tableau when none is available by inspection.

*Method of finding a b.f.s.*

A basic non-negative solution of the equations

$$A\mathbf{x} = \mathbf{b} \qquad (9)$$

is found by first arranging the equations so that $\mathbf{b} \geq \mathbf{0}$ and then by solving the c.m.p.

minimise $\mathbf{u}'\mathbf{w}$,

subject to $A\mathbf{x} + \mathbf{w} = \mathbf{b}$ and $\mathbf{x} \geq \mathbf{0}, \mathbf{w} \geq \mathbf{0}$,

where the $m$-vector $\mathbf{u} = (1, 1, \ldots, 1)'$, so that $\mathbf{u}'\mathbf{w} = \sum_{i=1}^{m} w_i$. An immediate b.f.s. is $\mathbf{x} = \mathbf{0}$, $\mathbf{w} = \mathbf{b} \geq \mathbf{0}$ and the corresponding basis vectors are the unit vectors $\mathbf{e}_1, \mathbf{e}_2, \ldots, \mathbf{e}_m$, as $\mathbf{w} = \sum_{i=1}^{m} w_i\mathbf{e}_i$. The problem has an optimal solution as $\mathbf{u}'\mathbf{w} \geq 0$ for all feasible $\mathbf{w}$, and it is easy to see that (9) has a non-negative solution if and only if the minimum value of $\mathbf{u}'\mathbf{w}$ is zero.

*Example*

Minimise $\quad x_1 + x_2 + 3x_3 + x_4$

subject to $\quad 2x_1 + 3x_2 \qquad + 6x_4 \geq 8$

$\qquad\qquad\ \ 3x_1 + x_2 + 2x_3 - 7x_4 = -4 \qquad (10)$

and $\qquad x_i \geq 0.$

*Solution.* The problem is equivalent to the following c.m.p., where $\mathbf{b} \geq \mathbf{0}$.

Minimise $\quad x_1 + x_2 + 3x_3 + x_4,$

subject to $\quad 2x_1 + 3x_2 \qquad + 6x_4 - x_5 = 8,$

$\qquad\qquad -3x_1 - x_2 - 2x_3 + 7x_4 \qquad = 4, \qquad (11)$

and $\qquad x_i \geq 0.$

To find a b.f.s., we solve the c.m.p.

minimise $\qquad w_1 + w_2,$

subject to $\qquad 2x_1 + 3x_2 \qquad + 6x_4 - x_5 + w_1 \qquad = 8,$

$\qquad\qquad\qquad -3x_1 - x_2 - 2x_3 + 7x_4 \qquad\qquad + w_2 = 4,$

and $\qquad x_i \geq 0, \ w_i \geq 0.$

for which a b.f.s. is $\mathbf{x} = \mathbf{0}$ and $\mathbf{w} = (8, 4)'$ and the basis vectors are $\mathbf{e}_1$ and $\mathbf{e}_2$. We now solve the problem by the simplex method.

| | c | 0 | 0 | 0 | 0 | 0 | 1 | 1 | |
|---|---|---|---|---|---|---|---|---|---|
| | | $\mathbf{a}_1$ | $\mathbf{a}_2$ | $\mathbf{a}_3$ | $\mathbf{a}_4$ | $\mathbf{a}_5$ | $\mathbf{e}_1$ | $\mathbf{e}_2$ | $\mathbf{b}$ |
| 1 | $\mathbf{e}_1$ | 2 | 3 | 0 | 6 | −1 | 1 | 0 | 8 |
| 1 | $\mathbf{e}_2$ | −3 | −1 | −2 | $\boxed{7}$ | 0 | 0 | 1 | 4 |
| | $\mathbf{z} - \mathbf{c}$ | −1 | 2 | −2 | 13 | −1 | 0 | 0 | 12 |
| | $\mathbf{e}_1$ | $\boxed{\dfrac{32}{7}}$ | $\dfrac{27}{7}$ | $\dfrac{12}{7}$ | 0 | −1 | 1 | $-\dfrac{6}{7}$ | $\dfrac{32}{7}$ |
| | $\mathbf{a}_4$ | $-\dfrac{3}{7}$ | $-\dfrac{1}{7}$ | $-\dfrac{2}{7}$ | 1 | 0 | 0 | $\dfrac{1}{7}$ | $\dfrac{4}{7}$ |
| | $\mathbf{z} - \mathbf{c}$ | $\dfrac{32}{7}$ | $\dfrac{27}{7}$ | $\dfrac{12}{7}$ | 0 | −1 | 0 | $-\dfrac{13}{7}$ | $\dfrac{32}{7}$ |
| | $\mathbf{a}_1$ | 1 | $\dfrac{27}{32}$ | $\dfrac{3}{8}$ | 0 | $-\dfrac{7}{32}$ | $\dfrac{7}{32}$ | $-\dfrac{3}{16}$ | 1 |
| | $\mathbf{a}_4$ | 0 | $\dfrac{7}{32}$ | $-\dfrac{1}{8}$ | 1 | $-\dfrac{3}{32}$ | $\dfrac{3}{32}$ | $\dfrac{1}{16}$ | 1 |
| | $\mathbf{z} - \mathbf{c}$ | 0 | 0 | 0 | 0 | 0 | −1 | −1 | 0 |

From the above tableau, a b.f.s. for (11) is $\mathbf{x} = (1, 0, 0, 1, 0)'$, so we now calculate the $\mathbf{z} - \mathbf{c}$ row appropriate to problem (11) and proceed with the simplex method. The vectors $\mathbf{e}_1$ and $\mathbf{e}_2$ are retained, as we shall show how to obtain an optimal solution to the dual problem from the optimal simplex tableau.

| c | | 1 | 1 | 3 | 1 | 0 | 0 | 0 | |
|---|---|---|---|---|---|---|---|---|---|
| | | $a_1$ | $a_2$ | $a_3$ | $a_4$ | $a_5$ | $e_1$ | $e_2$ | b |
| 1 | $a_1$ | 1 | $\boxed{\dfrac{27}{32}}$ | $\dfrac{3}{8}$ | 0 | $-\dfrac{7}{32}$ | $\dfrac{7}{32}$ | $-\dfrac{3}{16}$ | 1 |
| 1 | $a_4$ | 0 | $\dfrac{7}{32}$ | $-\dfrac{1}{8}$ | 1 | $-\dfrac{3}{32}$ | $\dfrac{3}{32}$ | $\dfrac{1}{16}$ | 1 |
| | $z-c$ | 0 | $\dfrac{1}{16}$ | $-\dfrac{11}{4}$ | 0 | $-\dfrac{5}{16}$ | $\dfrac{5}{16}$ | $-\dfrac{1}{8}$ | 2 |
| | $a_2$ | | | | | | | | $\dfrac{32}{27}$ |
| | $a_4$ | | | | | | | | $\dfrac{20}{27}$ |
| | $z-c$ | $-\dfrac{2}{27}$ | 0 | $-\dfrac{25}{9}$ | 0 | $-\dfrac{8}{27}$ | $\dfrac{8}{27}$ | $-\dfrac{1}{9}$ | $\dfrac{52}{27}$ |

Since $z_j - c_j \leq 0$ for $j = 1, 2, \ldots, 5$, the optimal solution is $\mathbf{x} = \left(0, \dfrac{32}{27}, 0, \dfrac{20}{27}, 0\right)'$ and the value of the programme is $\dfrac{52}{27}$.

The dual of the problem (11) is to

maximise $8y_1 + 4y_2$

subject to $2y_1 - 3y_2 \leq 1$, $3y_1 - y_2 \leq 1$, $-2y_2 \leq 3$, $6y_1 + 7y_2 \leq 1$,

$-y_1 \leq 0$.

It is easy to verify that the numbers in the bottom row of the tableau under $e_1$ and $e_2$, $\mathbf{y} = \left(\dfrac{8}{27}, -\dfrac{1}{9}\right)'$, are feasible for the dual problem and the optimality of $\mathbf{y}$ follows as $8y_1 + 4y_2 = \dfrac{52}{27}$, the value of the programme. We now show that the simultaneous solution of the dual problem was no accident.

61

## Solution of the dual problem

Suppose that we have an optimal solution for the c.m.p. and that the corresponding simplex tableau is as shown below, where $e_j = \sum_{i=1}^{m} l_{ij}\mathbf{a}_i$ and $y_j = \sum_{i=1}^{m} l_{ij}c_i$.

| | **c** | $c_1$ $c_m$ | $c_{m+1}$ | $c_n$ | 0 | 0 | |
|---|---|---|---|---|---|---|---|
| | | $\mathbf{a}_1$ $\mathbf{a}_m$ | $\mathbf{a}_{m+1}$ | $\mathbf{a}_n$ | $\mathbf{e}_1$ | $\mathbf{e}_m$ | **b** |
| $c_1$ | $\mathbf{a}_1$ | 1  0 | $t_{1m+1}$ | $t_{1n}$ | $l_{11}$ | $l_{1m}$ | $x_1$ |
| $c_m$ | $\mathbf{a}_m$ | 0  1 | $t_{mm+1}$ | $t_{mn}$ | $l_{m1}$ | $l_{mm}$ | $x_m$ |
| **z − c** | | 0  0 | $z_{m+1}-c_{m+1}$ | $z_n-c_n$ | $y_1$ | $y_m$ | $z_0$ |

Let $A_b = (\mathbf{a}_1, \mathbf{a}_2, \ldots, \mathbf{a}_m)$ be the matrix of the basis vectors and $L = (l_{ij})$, then it follows from (2) that the unit matrix $I$ satisfies

$$I = (\mathbf{e}_1, \mathbf{e}_2, \ldots, \mathbf{e}_m) = A_b L,$$

and since $A_b$ is regular,

$$LA_b = I, \tag{12}$$

so that $L$ is the inverse basis matrix. On writing

$$\mathbf{y}' = (y_j) = \mathbf{c}_b' L \quad \text{where} \quad \mathbf{c}_b = (c_1, c_2, \ldots, c_m)',$$

it follows from (12) that

$$\mathbf{y}'A_b = \mathbf{c}_b' \quad \text{or} \quad \mathbf{y}'\mathbf{a}_i = c_i \quad \text{for} \quad i = 1, 2, \ldots, m.$$

For $j = m + 1, \ldots, n$

$$\mathbf{y}'\mathbf{a}_j = \sum_{i=1}^{m} \mathbf{y}'t_{ij}\mathbf{a}_i = \sum_{i=1}^{m} t_{ij}c_i = z_j \leq c_j, \tag{13}$$

since the optimality criterion is satisfied. Hence $\mathbf{y}$ satisfies the dual constraints, $\mathbf{y}'A \leq \mathbf{c}'$, and since $x_i > 0 \Rightarrow \mathbf{y}'\mathbf{a}_i = c_i$, $\mathbf{y}$ is optimal by Theorem 5 on page 23. It can be shown that the optimal vector for the dual problem is unique if the primal (original) problem is non-degenerate (see exercise 4 on page 66).

We have now shown how to solve any linear programming problem. The procedure is to first transform the problem into a c.m.p., if necessary, then to use the technique described on page 58 to find a b.f.s. and the corresponding simplex tableau, if one is not available by inspection, and finally to solve the problem. Degeneracy is the only outstanding problem and we now show how to overcome this difficulty, though it is usually worth trying another replacement to see if the value of $z_0$ can be reduced.

## Degeneracy

The simplex tableau on page 62, extended by the unit vectors, $e_1, e_2, \ldots, e_m$, is in a suitable form for dealing with the degeneracy problem. The degenerate constraint equation, $A\mathbf{x} = \mathbf{b}$, is made non-degenerate by adding the $m$-vector, $(\varepsilon, \varepsilon^2, \ldots, \varepsilon^m)' = \sum_{j=1}^{m} \varepsilon^j e_j$, to $\mathbf{b}$, where $\varepsilon$ is a suitably small number. Suppose that during the course of solving the unperturbed problem, the $\min_{t_{is}>0} \dfrac{x_i}{t_{is}}$ is not unique and that $t_{rs}$ and $t_{qs}$ (say) both qualify for the pivot. If $t_{rs}$ is chosen, then the subsequent replacement will result in a zero value for $x_q^*$ and the solution will be degenerate. However in the perturbed problem, the pivot is uniquely determined by the

$$\min_{t_{is}>0} \frac{1}{t_{is}} \left( x_i + \sum_{j=1}^{m} \varepsilon^j l_{ij} \right), \tag{14}$$

since it follows from the tableau that

$$\mathbf{b} + \sum_{j=1}^{m} \varepsilon^j e_j = \sum_{i=1}^{m} \left( x_i + \sum_{j=1}^{m} \varepsilon^j l_{ij} \right) \mathbf{a}_i.$$

Now $\varepsilon$ is arbitrarily small, so the minimum will occur at either $i = r$ or $i = q$ and to determine which will give the pivot, we compare the coefficients of the powers of $\varepsilon$ in (14) for $i = r$ and $i = q$. This leads to a comparison of the vectors

$$\frac{1}{t_{rs}}(l_{r1}, l_{r2}, \ldots l_{rm})' \quad \text{and} \quad \frac{1}{t_{qs}}(l_{q1}, l_{q2}, \ldots, l_{qm})'.$$

If the above vectors first differ in the $k$th element and

$$\frac{l_{rk}}{t_{rs}} < \frac{l_{qk}}{t_{qs}},$$

then $t_{rs}$ is the pivot for the replacement, as the higher powers of $\varepsilon$ can be neglected. The method can easily be extended to cover the case in which more than two $t_{is}$ qualify for the pivot.

We conclude this chapter by describing briefly two variations of the simplex method. For a more detailed account of the methods, the reader is referred to the books by Krekó and Beale, which are listed under the 'Suggestions for Further Reading' on page 83.

## Inverse Matrix Method (Revised Simplex Method)

This method consists of only up-dating the inverse basis matrix $L$ and the values of $\mathbf{x}$, $\mathbf{y}$ and $z_0$ (see the tableau on page 62). The values of $z_j - c_j$ are calculated from the matrix $A$ using the relation (13), namely

$$z_j - c_j = \mathbf{y}'\mathbf{a}_j - c_j,$$

and the pivotal column is chosen, as in the simplex method, by the criterion $z_s - c_s > 0$. Then the elements of the pivotal column are calculated from the equation

$$t_{is} = (LA)_{is},$$

which follows from (2), and the pivot is given by the $\min_{t_{is} > 0} \dfrac{x_i}{t_{is}}$ .

The inverse matrix method is used in most computer programmes as it involves less computations when the number of variables is large compared with the number of constraints and when the matrix $A$ contains several zeros. It is also a better method for handling degeneracy and for doing sensitivity analysis, which gives the criteria under which $\mathbf{a}_1, \mathbf{a}_2, \ldots, \mathbf{a}_m$ remain an optimal basis for changes in the values of $\mathbf{b}, \mathbf{c}$ and $A$. (see exercises 8 and 9 on page 67).

## Dual Simplex Method

The dual simplex method solves the c.m.p. by using a tableau for which the optimality criterion, $z_j \leq c_j$, is satisfied for all replacements, but the corresponding $x_i$'s are not necessarily non-negative. If all the $x_i$'s are non-negative, then the solution is optimal by Theorem 2. If not, the pivot for the next replacement is obtained by choosing some $x_r < 0$ and calculating the $\min\limits_{t_{rj} < 0} \dfrac{z_j - c_j}{t_{rj}} = \dfrac{z_s - c_s}{t_{rs}}$ to give the pivot $t_{rs}$. In this way $x_r^* = \dfrac{x_r}{t_{rs}} > 0$ and $z_j^* - c_j \leq 0$. The value of $z_0$ in the dual simplex method is the current value of the dual problem, which is to maximise $\mathbf{y}'\mathbf{b}$ subject to $\mathbf{y}'A \leq \mathbf{c}'$, and the replacement increases the value of $z_0$ by $-\dfrac{(z_s - c_s)x_r}{t_{rs}} \geq 0$.

The dual simplex method is applicable to the following type of c.m.p:

minimise $\mathbf{c}'\mathbf{x}$,

subject to $A\mathbf{x} + \mathbf{w} = \mathbf{b}$ and $\mathbf{x} \geq 0$, $\mathbf{w} \leq 0$,

where $\mathbf{c} \geq 0$ and $\mathbf{b} \not\geq 0$. For this problem $\mathbf{w} = \mathbf{b}$ is a basic non-feasible solution of $A\mathbf{x} + \mathbf{w} = \mathbf{b}$ and the corresponding

$z_j - c_j = - c_j \le 0$ for $j = 1, 2, \ldots, n$. The method is also useful when an additional constraint is to be added to the simplex tableau.

## EXERCISES ON CHAPTER FOUR

1. Solve by the simplex method:

minimise $\qquad x_1 + 6x_2 + 2x_3 - x_4 + x_5 - 3x_6$,

subject to $\qquad$
$$x_1 + 2x_2 + x_3 \qquad\qquad + 5x_6 = 3,$$
$$-3x_2 + 2x_3 + x_4 \qquad + x_6 = 1,$$
$$5x_2 + 3x_3 \qquad + x_5 - 2x_6 = 2$$

and $\qquad x_i \ge 0$.

2. Solve by the simplex method:

minimise $\qquad 4x_1 + x_2 - x_3 + 2x_4$,

subject to $\qquad$
$$3x_1 - 3x_2 + x_3 \qquad\quad = 3,$$
$$6x_2 - 2x_3 + x_4 = 2$$

and $\qquad x_i \ge 0$.

3. If $\mathbf{a}_1, \mathbf{a}_2, \ldots, \mathbf{a}_m$ is an optimal basis for a c.m.p. and if

$$z_j < c_j \text{ for } j = m + 1, \ldots, n,$$

show that the c.m.p. has a unique optimal vector.

4. Show that for any non-degenerate c.m.p., the optimal vector for the dual problem is unique, if it exists, and that the optimality criterion on page 53 is satisfied for all optimal bases.

5. Solve by the simplex method for all values of $\lambda$:

minimise $\qquad \lambda x_1 + 2x_2 - 5x_3 + x_4$,

subject to $\qquad$
$$2x_1 + x_2 - 3x_3 - x_4 = 3,$$
$$3x_1 - 4x_2 + x_3 + 6x_4 = 1,$$
$$-x_1 - x_2 + 2x_3 + 2x_4 = -1$$

and $\qquad x_i \ge 0$.

(This is an example of parametric linear programming.)

6. Solve by the simplex method:

maximise $\qquad y_1 + 4y_2 + 3y_3,$

subject to $\qquad 3y_1 + 2y_2 + \ y_3 \leq \ 4,$

$\qquad\qquad\quad y_1 + 5y_2 + 4y_3 \leq 14,$

and $\qquad\qquad y_i \geq 0.$

Obtain the solution to the dual problem from the simplex tableau.

7. Solve by the simplex method:

maximise $\qquad x_1 + 6x_2 + 4x_3,$

subject to $\qquad -x_1 + 2x_2 + 2x_3 \leq 13,$

$\qquad\qquad\quad 4x_1 - 4x_2 + \ x_3 \leq 20,$

$\qquad\qquad\quad x_1 + 2x_2 + \ x_3 \leq 17,$

$\qquad\qquad\quad x_1 \leq 7, \ x_2 \geq 2, \ x_3 \geq 3.$

8. A baby cereal is to be marketed which contains 25% protein and not more than 70% carbohydrate. The manufacturer has four cereals at his disposal, whose analysis and cost are given in the table below.

| Cereal | $C_1$ | $C_2$ | $C_3$ | $C_4$ |
|---|---|---|---|---|
| % carbohydrate | 80 | 80 | 70 | 60 |
| % protein | 10 | 15 | 20 | 30 |
| Cost per pound in shillings | 1 | 2 | 3 | 4 |

What blend of the four cereals should the manufacturer use to minimise the cost of the baby cereal? By how much must the cost of $C_2$ be reduced so that $C_2$ is as cheap to use as $C_1$ in the manufacture of the baby cereal, and give the blend in this case?

9. Show that an optimal basis for the c.m.p. remains optimal under the perturbation $\mathbf{b} \to \mathbf{b} + \delta\mathbf{b}$, provided that $L(\mathbf{b} + \delta\mathbf{b}) \geq \mathbf{0}$, where $L$ is the inverse basis matrix.

# CHAPTER FIVE

# Game Theory

## 1. Two-person Zero-sum Games

Game theory models competitive situations and has applications in management science, military tactics and economic theory. In this chapter the discussion will be confined to two-person zero-sum games. These are games between two players, which result in one player paying the other a predetermined amount, called the payoff. The underlying assumption of the theory is that both players try to win, and this rules out any cooperation between the players. We use the following simple card game to illustrate the concepts of the theory.

Two players, $P_1$ and $P_2$, are each dealt a card from a pack of three cards. After looking at his card, $P_1$ guesses which card remains in the pack. $P_2$, having heard $P_1$'s guess and having inspected his own card, also guesses the identity of the third card. If either player guesses correctly, he wins 1 from his opponent.

The first step in analysing the game is to determine the possible strategies for $P_1$ and $P_2$. A strategy is a set of instructions determining the player's choice at every move in the game and in all conceivable situations. Even though a player may not have thought out a course of action, he will behave as though he had a strategy. $P_1$ has two strategies which are $s_1$: guess his own card and $s_2$: guess one of the other two cards with equal probability. $P_2$'s strategies depend upon $P_1$'s move and are given in the table below, where those strategies which involve $P_2$ choosing his own card, have been eliminated as $P_2$ can clearly do better by a choice of one of the other two cards.

| $P_2$'s strategies | $P_1$ guesses $P_2$'s card | $P_1$ doesn't guess $P_2$'s card |
|---|---|---|
| $t_1$ | Guess one of the other two cards with equal probability. | Guess the same as $P_1$. |
| $t_2$ | ditto | Guess the remaining card. |

Let $\phi(s_i, t_j)$ be the payoff to $P_1$ as a result of the game in which $P_1$ and $P_2$ use $s_i$ and $t_j$ respectively, then $-\phi(s_i, t_j)$ is the payoff to $P_2$. If $P_1$ plays $s_1$ and $P_2$ plays $t_1$, then both guess $P_1$'s card and win nothing, so $\phi(s_1, t_1) = 0$. If $P_1$ plays $s_2$ and $P_2$ plays $t_2$, then either $P_1$ guesses $P_2$'s card and $P_2$ guesses the card in the pack with probability $\frac{1}{2}$ or $P_1$ guesses the card in the pack and $P_2$ guesses $P_1$'s card, so the expected (average) payoff to $P_1$ is $\phi(s_2, t_2) = \frac{1}{2} \times \frac{1}{2} \times (-1) + \frac{1}{2} \times 1$. The remaining payoffs, $\phi(s_1, t_2)$ and $\phi(s_2, t_1)$, are calculated similarly and the payoff matrix, with $(i,j)$th element $\phi(s_i, t_j)$, is given by (1).

$$\begin{array}{cc} & \begin{array}{cc} t_1 & t_2 \end{array} \\ \begin{array}{c} s_1 \\ s_2 \end{array} & \left( \begin{array}{cc} 0 & -1 \\ -\frac{1}{4} & \frac{1}{4} \end{array} \right) \end{array} \qquad (1)$$

In this game there is no single best strategy for either $P_1$ or $P_2$ for if $P_1$ always uses $s_2$, $P_2$ will use $t_1$, in which case it would be better for $P_1$ to use $s_1$ etc. It will be shown later that a best policy for both players is to vary their choice of strategy in a sequence of games, that is it pays $P_1$ to bluff occasionally.

DEFINITION. A two-person zero-sum game consists of two strategy sets $S$ and $T$, where $s \in S$ and $t \in T$ are the strategies for the players $P_1$ and $P_2$ respectively. The payoff, $\phi(s,t)$, represents the amount paid by $P_1$ to $P_2$ as a result of the game in which $P_1$ plays $s$ and $P_2$ plays $t$, and is a real valued function defined for all $s \in S$ and $t \in T$.

The process of determining $S, T$ and $\phi(s,t)$ for a game is called normalisation. If $s$ or $t$ involve chance moves, then $\phi(s,t)$ is the expected payoff. In the cases when $S$ and $T$ are finite the payoff can be represented as a matrix, as in (1).

## 2. Solution of Games: Saddle Points

We now consider the game with payoff matrix (2). Since $P_2$

$$
\begin{array}{c}
\begin{array}{ccc}
t_1 & t_2 & t_3
\end{array} \\
\begin{array}{c}
s_1 \\
s_2 \\
s_3
\end{array}
\left(
\begin{array}{ccc}
3 & 0 & -1 \\
-1 & 0 & 3 \\
2 & 1 & 2
\end{array}
\right)
\end{array}
\qquad (2)
$$

will try to minimise $P_1$'s gains, a sensible procedure for selecting $P_1$'s best strategy is to determine the minimum payoff for each of his strategies, that is $-1$ for $s_1$, $-1$ for $s_2$ and $1$ for $s_3$ and then to select the strategy corresponding to the maximum of these payoffs (maximin $\phi$), which is $s_3$. The similar procedure for $P_2$ is to find the most he can lose by using each of his strategies, that is 3 for $t_1$, 1 for $t_2$ and 3 for $t_3$, and then to select the strategy corresponding to the minimum of these payoffs (minimax $\phi$), which is $t_2$. In this game $P_1$ and $P_2$ will be satisfied with their maximin and minimax strategies, $s_3$ and $t_2$ respectively, for

$$
\phi(s_3,t_1), \; \phi(s_3,t_3) \geq \phi(s_3,t_2) \geq \phi(s_1,t_2), \; \phi(s_2,t_2), \qquad (3)
$$

which means that neither player could have improved their choice of strategy had they known their opponents choice. This reasoning motivates the following definition.

DEFINITION. The game $G = (S,T;\phi)$ has the solution $(\bar{s},\bar{t};\omega)$ if

$$
\phi(\bar{s},t) \geq \omega \geq \phi(s,\bar{t}) \text{ for all } s \in S \text{ and } t \in T,
$$

where $\omega = \phi(\bar{s},\bar{t})$ is called the value of the game and $\bar{s} \in S$, $\bar{t} \in T$ are called the optimal strategies.

A function $\phi(s,t)$, defined for $s \in S$ and $t \in T$, is said to have a saddle point at $(\bar{s},\bar{t})$ if

$$\phi(\bar{s},t) \geq \phi(s,\bar{t}) \text{ for all } s \in S \text{ and } t \in T,$$

so if the payoff has a saddle point at $(\bar{s},\bar{t})$, then $(\bar{s},\bar{t};\phi(\bar{s},\bar{t}))$ is a solution to the game. If the payoff is representable by a matrix, then a matrix element is a saddle point if it is the minimum in its row and the maximum in its column.

DEFINITION. For the game $G = (S,T;\phi)$,

$$\text{maximin } \phi = \max_{s \in S} \min_{t \in T} \phi(s,t) \text{ and minimax } \phi = \min_{t \in T} \max_{s \in S} \phi(s,t)$$

provided they exist.

By using his maximin strategy $P_1$ is assured of winning at least maximin $\phi$ and by using his minimax strategy $P_2$ will not lose more than minimax $\phi$. For the payoff matrix (2) the maximin and minimax are equal and it follows from (3) that $(s_3,t_2;1)$ is a solution to the game. However for the game with payoff matrix (1), maximin $\phi = -\frac{1}{4} <$ minimax $\phi = 0$ and there was no single best strategy for either player. The following theorem gives a criterion for the existence of a saddle point.

THEOREM 1. If maximin $\phi$ and minimax $\phi$ exist, then

$$\text{maximin } \phi \leq \text{minimax } \phi,$$

and $\phi$ has a saddle point if and only if

$$\text{maximin } \phi = \text{minimax } \phi.$$

*Proof.* Since

$$\min_{t \in T} \phi(s,t) \leq \phi(s,t) \leq \max_{s \in S} \phi(s,t) \text{ for all } s \in S \text{ and } t \in T, \quad (4)$$

$$\text{maximin } \phi \leq \text{minimax } \phi.$$

71

Suppose maximin $\phi$ = minimax $\phi$, then there exist $\bar{s} \in S$ and $\bar{t} \in T$ such that

$$\min_{t \in T} \phi(\bar{s},t) = \text{maximin } \phi = \text{minimax } \phi = \max_{s \in S} \phi(s,\bar{t}).$$

It now follows from (4) that

$$\phi(\bar{s},t) \geq \phi(s,\bar{t}) \text{ for all } s \in S \text{ and } t \in T,$$

so $\phi$ has a saddle point at $(\bar{s},\bar{t})$. Conversely suppose that $\phi$ has a saddle point at $(\bar{s},\bar{t})$, then

$$\phi(\bar{s},t) \geq \phi(s,\bar{t}) \text{ for all } s \in S \text{ and } t \in T,$$

so

$$\min_{t \in T} \phi(\bar{s},t) \geq \max_{s \in S} \phi(s,\bar{t})$$

and maximin $\phi \geq$ minimax $\phi$. Hence by the first part of the theorem, equality must hold.

## 3. Solution of Games: Mixed Strategies

If the payoff has no saddle point then the game has no solution in pure strategies, that is in strategies belonging to the strategy sets $S$ and $T$. So to make progress we must introduce the mixed strategy, which is represented by a probability density function defined on a strategy set.

DEFINITION. The mixed strategies $x(s)$ and $y(t)$ for $P_1$ and $P_2$ respectively, are real valued functions satisfying

$$\sum_{s \in S} x(s) = 1 \text{ and } x(s) \geq 0 \text{ for all } s \in S, \tag{5}$$

$$\sum_{t \in T} y(t) = 1 \text{ and } y(t) \geq 0 \text{ for all } t \in T.$$

The corresponding value of the payoff is given by the expected payoff

$$\phi(x, y) = \sum_{s \in S} \sum_{t \in T} x(s) \, y(t) \, \phi(s,t) \, .$$

If $P_1$ plays a mixed strategy, then he must use a probabilistic device, such as a dice or pack of cards, to determine which pure strategy to use in each play of the game. A pure strategy, $s_i$ say, can be thought of as the mixed strategy satisfying $x(s_i) = 1$ and $x(s) = 0$ for $s \neq s_i$. We now extend the definition of the solution to a game to include mixed strategies.

DEFINITION. The game $G = (S, T; \phi)$ has the solution $(\bar{x}(s), \bar{y}(t); \omega)$ if

$$\phi(\bar{x},y) \geq \omega \geq \phi(x,\bar{y}) \quad \text{for all mixed strategies } x(s) \text{ and } y(t) \quad (6)$$

or equivalently

$$\phi(\bar{x},t) \geq \omega \geq \phi(s,\bar{y}) \quad \text{for all } s \in S \text{ and } t \in T, \quad (7)$$

where $\omega = \phi(\bar{x},\bar{y})$ is the value of the game and $\bar{x}(s)$ and $\bar{y}(t)$ are called optimal strategies.

To verify the equivalence of (6) and (7), we assume that (7) holds, and then use (5) to show that

$$\phi(\bar{x},y) = \sum_{t \in T} y(t)\phi(\bar{x},t) \geq \sum_{t \in T} y(t)\omega = \omega \quad \text{for all mixed } y(t).$$

Similarly $\phi(x,\bar{y}) \leq \omega$ for all mixed $x(s)$. Condition (7) follows from (6), as pure strategies are special cases of mixed strategies.

The condition (7) is easier to apply than (6) and as an example, we show that $\left(\bar{x}, \bar{y}; -\dfrac{1}{6}\right)$ is a solution to the game with payoff matrix (1), where $\bar{x}(s_1) = \dfrac{1}{3}$ , $\bar{x}(s_2) = \dfrac{2}{3}$ , $\bar{y}(t_1) = \dfrac{5}{6}$ and

73

$\bar{y}(t_2) = \dfrac{1}{6}$. On applying (7) we obtain

$$\phi(\bar{x}, t_1) = \frac{1}{3} \times 0 + \frac{2}{3} \times \left(-\frac{1}{4}\right) = -\frac{1}{6}, \qquad \phi(\bar{x}, t_2) = -\frac{1}{6},$$

$$\phi(s_1, \bar{y}) = \frac{5}{6} \times 0 + \frac{1}{6} \times (-1) = -\frac{1}{6}, \qquad \phi(s_2, \bar{y}) = -\frac{1}{6},$$

so $\phi(\bar{x}, t) \geq -\dfrac{1}{6} \geq \phi(s, \bar{y})$ for all $s \in S$ and $t \in T$. An interpretation of (7) is that an optimal strategy for $P_1$ achieves at least the value of the game against all the pure strategies of $P_2$.

In the next theorem we prove that the value of a game is unique, though the reader should note that the optimal strategies are not necessarily unique.

THEOREM 2. The value of a game, if it exists, is unique.

*Proof.* Suppose $(\bar{x}, \bar{y}; \omega)$ and $(x^*, y^*; \omega^*)$ are solutions of the game $G = (S, T; \phi)$, then it follows from (6) that

$$\omega = \phi(\bar{x}, \bar{y}) \geq \phi(x^*, \bar{y}) \geq \phi(x^*, y^*) = \omega^*$$

and

$$\omega = \phi(\bar{x}, \bar{y}) \leq \phi(\bar{x}, y^*) \leq \phi(x^*, y^*) = \omega^*,$$

which proves that $\omega = \omega^*$.


## 4. Dominated and Essential Strategies

A strategy $s_k$ for $P_1$ is said to dominate his strategy $s_l$ if for all $t \in T$, $\phi(s_k, t) \geq \phi(s_l, t)$, which says that $P_1$ will not be worse off by using $s_k$ in preference to $s_l$, whatever strategy $P_2$ employs. Similarly the strategy $t_k$ for $P_2$ dominates his strategy $t_l$ if $\phi(s, t_k) \leq \phi(s, t_l)$ for all $s \in S$. The following theorem proves that if a dominated strategy is eliminated from a game, then any solution to the reduced game is a solution to the original game.

THEOREM 3. If $\phi(s_k,t) \geq \phi(s_l,t)$ for all $t \in T$ and if $(x^*,\bar{y};\omega)$ is a solution to the game $G^* = (S - s_l, T; \phi)$, then $(\bar{x},\bar{y};\omega)$ is a solution to the game $G = (S,T;\phi)$, where $\bar{x}(s) = x^*(s)$ for $s \in S - s_l$ and $\bar{x}(s_l) = 0$.

*Proof.* Since $(x^*,\bar{y};\omega)$ is a solution to $G^* = (S - s_l, T; \phi)$

$$\phi(x^*,t) \geq \omega \geq \phi(s,\bar{y}) \quad \text{for all } s \in S - s_l \text{ and } t \in T.$$

From the definition of $\bar{x}(s)$

$$\phi(\bar{x}, t) = \sum_{s \in S} \bar{x}(s)\, \phi(s,t) = \sum_{s \in S - s_l} x^*(s)\, \phi(s, t) = \phi(x^*, t) \geq \omega$$

and

$$\phi(s_l, \bar{y}) = \sum_{t \in T} \bar{y}(t)\, \phi(s_l, t) \leq \sum_{t \in T} \bar{y}(t)\, \phi(s_k, t) = \phi(s_k, \bar{y}) \leq \omega$$

Hence $\phi(\bar{x},t) \geq \omega \geq \phi(s,\bar{y})$ for all $s \in S$ and $t \in T$, and $(\bar{x},\bar{y};\omega)$ is a solution to $G = (S,T;\phi)$.

A pure strategy is called essential if it is used in some optimal strategy and in the next theorem, we prove that an essential pure strategy of one player achieves the value of the game against all the optimal strategies of his opponent.

DEFINITION. The pure strategy $s_k$ for $P_1$ is called essential if there exists an optimal strategy $\bar{x}(s)$ for $P_1$, with $\bar{x}(s_k) > 0$.

THEOREM 4. If the game $G = (S,T;\phi)$ has a solution and if $s_k$ is an essential pure strategy for $P_1$, then

$$\phi(s_k,\bar{y}) = \omega \quad \text{for all optimal strategies } \bar{y}(t) \text{ of } P_2,$$

where $\omega$ is the value of the game and the strategy sets, $S$ and $T$, are enumerable.

*Proof.* As $s_k$ is essential, there exists an optimal strategy, $\bar{x}(s)$ for $P_1$, with $\bar{x}(s_k) > 0$. For all optimal strategies $\bar{y}(t)$ of $P_2$, $\phi(\bar{x},\bar{y}) = \omega$, which can also be written as

$$\sum_{s \in S} \bar{x}(s)\big(\phi(s,\bar{y}) - \omega\big) = 0 . \tag{8}$$

Since $\bar{y}$ is optimal, $\phi(s,\bar{y}) \leq \omega$ for all $s \in S$ and by the definition of a mixed strategy, $\bar{x}(s) \geq 0$ for all $s \in S$, hence it follows from (8) that as $S$ is enumerable,

$$\bar{x}(s)(\phi(s,\bar{y}) - \omega) = 0 \text{ for all } s \in S,$$

which implies that $\phi(s_k,\bar{y}) = \omega$, as $\bar{x}(s_k) > 0$.

The following example illustrates how Theorems 3 and 4 can be used to solve a game.

*Example.* Find a solution to the game with payoff matrix

$$
\begin{array}{c}
\quad \\
s_1 \\
s_2 \\
s_3
\end{array}
\begin{array}{c}
\begin{array}{cccc}
t_1 & t_2 & t_3 & t_4
\end{array} \\
\left(
\begin{array}{cccc}
4 & 4 & 6 & 6 \\
6 & 5 & 4 & 4 \\
6 & 5 & 6 & 0
\end{array}
\right),
\end{array}
$$

*Solution.* Since maximin $\phi = 4$ and minimax $\phi = 5$, the payoff has no saddle point by Theorem 1. The strategies $t_1$ and $t_3$ are dominated by the strategies $t_2$ and $t_4$ respectively, and in the reduced game, the strategy $s_3$ is dominated by $s_2$. We now use Theorem 3 and solve the reduced game

$$
\begin{array}{c}
\quad \\
s_1 \\
s_2
\end{array}
\begin{array}{c}
\begin{array}{cc}
t_2 & t_4
\end{array} \\
\left(
\begin{array}{cc}
4 & 6 \\
5 & 4
\end{array}
\right).
\end{array}
$$

Since the $2 \times 2$ matrix has no saddle point, the pure strategies for both players are essential (this result only holds for $2 \times 2$ matrices). Let $\bar{x}$ and $\bar{y}$ be optimal strategies for $P_1$ and $P_2$ respectively, where $\bar{x}(s_1) = p$, $\bar{x}(s_2) = 1 - p$, $\bar{y}(t_2) = q$, $\bar{y}(t_4) = 1 - q$ and $0 < p, q < 1$. On applying Theorem 4 we obtain

$$\phi(\bar{x}, t_1) = 4p + 5(1 - p) = \omega$$
$$\phi(\bar{x}, t_2) = 6p + 4(1 - p) = \omega$$
$$\phi(s_1, \bar{y}) = 4q + 6(1 - q) = \omega$$

which have the solution $p = \dfrac{1}{3}$, $\omega = \dfrac{14}{3}$ and $q = \dfrac{2}{3}$. Hence by Theorem 3, a solution to the game is for $P_1$ to use his strategies $s_1$ and $s_2$ with probabilities $\dfrac{1}{3}$ and $\dfrac{2}{3}$ respectively, for $P_2$ to use his strategies $t_2$ and

$t_4$ with probabilities $\dfrac{2}{3}$ and $\dfrac{1}{3}$ respectively, and the value of the game is $\dfrac{14}{3}$. The reader should check that the solution satisfies the condition (7).

## 5. Minimax Theorem

The existence of a solution to a game, $G = (S,T;\phi)$, can be proved under certain conditions, and the major result of this chapter will be to prove that $G$ has a solution when $S$ and $T$ are finite. Such a game is called a matrix game, as the payoff can be represented by a matrix. Existence can be proved for more general conditions.

### Matrix Games

If $S = (s_1, s_2, \ldots, s_m)$ and $T = (t_1, t_2 \ldots, t_n)$, then the mixed strategies $x(s)$ and $y(t)$ for $P_1$ and $P_2$ respectively, can be represented by the vectors $\mathbf{x} = (x_1, x_2, \ldots, x_m)'$ and $\mathbf{y} = (y_1, y_2, \ldots, y_n)'$, where $x_i = x(s_i)$ and $y_i = y(t_i)$. The conditions (5) for a mixed strategy are equivalent to

$$\mathbf{x}'\mathbf{u} = 1, \quad \mathbf{x} \geq \mathbf{0}, \quad \text{where } \mathbf{u} = (1, 1, \ldots, 1)'_m \qquad (9)$$

and $\quad \mathbf{v}'\mathbf{y} = 1, \quad \mathbf{y} \geq \mathbf{0}, \quad \text{where } \mathbf{v} = (1, 1, \ldots, 1)'_n. \qquad (10)$

On writing $\phi(s_i,t_j) = a_{ij}$, the expected payoff

$$\phi(x, y) = \sum_{i=1}^{m} \sum_{j=1}^{n} x_i y_j a_{ij} = \mathbf{x}'A\mathbf{y}, \qquad \text{where } A = (a_{ij})$$

and the condition (7) for $(\bar{\mathbf{x}},\bar{\mathbf{y}};\omega)$ to be a solution to the game

$G = (S,T;\phi)$ becomes

$$\bar{\mathbf{x}}'A \geq \omega\mathbf{v}', \qquad\qquad A\bar{\mathbf{y}} \leq \omega\mathbf{u}, \qquad (11)$$

as the pure strategies correspond to the unit vectors $\mathbf{e}_i$.

Before proving the existence theorem, we require the result that the sets of optimal strategies are unaltered by adding an arbitrary constant to the payoff, that is if $(x,y;\omega)$ is a solution to the game $G = (S,T;\phi)$, then $(x,y;\omega + k)$ is a solution to the game $G = (S,T,\phi + k)$ (proof left as an exercise for the reader).

THEOREM 5 (minimax theorem). A matrix game has a solution.

*Proof.* We assume that the maximin of the payoff matrix $A$ is positive, as this ensures that the value of the game is positive. If necessary, this can be achieved by adding a suitable constant to the elements of $A$. We now consider the s.m.p.,

minimise $\qquad \mathbf{x}'\mathbf{u}$

subject to $\qquad \mathbf{x}'A \geq \mathbf{v}'$ and $\mathbf{x} \geq \mathbf{0}$,

where $\mathbf{u} = (1, 1, \ldots, 1)'_m$ and $\mathbf{v} = (1, 1, \ldots, 1)'_n$. A feasible vector is $\mathbf{x} = a_{kl}^{-1}\mathbf{e}_k$ where $a_{kl}$ is the maximin. The dual problem is to

maximise $\qquad \mathbf{v}'\mathbf{y}$

subject to $\qquad A\mathbf{y} \leq \mathbf{u}$ and $\mathbf{y} \geq \mathbf{0}$,

for which $\mathbf{y} = \mathbf{0}$ is a feasible vector. Hence by the fundamental duality theorem on page 19, there exist optimal vectors $\mathbf{x}_0$ and $\mathbf{y}_0$ satisfying

$$\mathbf{x}_0'\mathbf{u} = \mathbf{v}'\mathbf{y}_0 = z_0$$

where $z_0 > 0$, as $\mathbf{x}_0$ must have at least one positive element to satisfy the constraints. Let

$$\bar{\mathbf{x}} = z_0^{-1}\mathbf{x}_0, \quad \bar{\mathbf{y}} = z_0^{-1}\mathbf{y}^0, \quad \omega = z_0^{-1},$$

then $(\bar{\mathbf{x}},\bar{\mathbf{y}};\omega)$ is a solution to the game, since $\bar{\mathbf{x}}$, $\bar{\mathbf{y}}$ and $\omega$ satisfy the conditions (9), (10) and (11).

The minimax theorem was first proved by von Neumann in 1928 using a fixed point theorem, whereas the above proof was given by Dantzig in 1951. The name of the theorem derives from the condition

$$\max_{x} \min_{y} \phi(x,y) = \min_{y} \max_{x} \phi(x,y),$$

which is necessary and sufficient for the existence of a solution to a game. This result is the generalisation of Theorem 1 to include mixed strategies.

## 6. Solution of Matrix Games by Simplex Method

The above proof of the minimax theorem is constructive, as it gives a technique for finding a solution to the game with payoff matrix $A$, which is to solve the related linear programme,

maximise $\quad$ $\mathbf{v}'\mathbf{y}$

subject to $\quad$ $A\mathbf{y} \leq \mathbf{u}$ and $\quad$ $\mathbf{y} \geq \mathbf{0}$. $\hspace{2em}$ (12)

*Example.* Find a solution to the game with payoff matrix

$$A = \begin{pmatrix} \frac{1}{3} & 0 & 1 \\ 1 & 0 & 0 \\ \frac{1}{3} & 1 & -1 \end{pmatrix},$$

*Solution.* Since maximin $A = 0$, we add 1 to the elements of $A$ and then convert (12) into the following canonical maximum problem,

maximise $\quad$ $y_1 + y_2 + y_3$

subject to $\quad$ $\dfrac{4}{3}y_1 + y_2 + 2y_3 + w_1 \hspace{3em} = 1$

$\hspace{5.5em} 2y_1 + y_2 + y_3 \hspace{1.5em} + w_2 \hspace{1.5em} = 1$

$\hspace{5.5em} \dfrac{4}{3}y_1 + 2y_2 \hspace{5em} + w_3 = 1$

and $\quad$ $y_i \geq 0, \; w_i \geq 0.$

A b.f.s. is $\mathbf{y} = \mathbf{0}$, $\mathbf{w} = (1, 1, 1)'$ and the corresponding simplex tableau is shown below. Since $\mathbf{v}'\mathbf{y}$ is to be maximised, $z_s < c_s$ is the criterion for $\mathbf{a}_s$ to be included in the basis and $z_j \geq c_j$ for all $j$, is the criterion for optimality (see footnotes on pages 51 and 53).

| c | | 1 | 1 | 1 | 0 | 0 | 0 | |
|---|---|---|---|---|---|---|---|---|
| | | $\mathbf{a}_1$ | $\mathbf{a}_2$ | $\mathbf{a}_3$ | $\mathbf{e}_1$ | $\mathbf{e}_2$ | $\mathbf{e}_3$ | $\mathbf{u}$ |
| 0 | $\mathbf{e}_1$ | $\frac{4}{3}$ | 1 | $\underline{2}$ | 1 | 0 | 0 | 1 |
| 0 | $\mathbf{e}_2$ | 2 | 1 | 1 | 0 | 1 | 0 | 1 |
| 0 | $\mathbf{e}_3$ | $\frac{4}{3}$ | 2 | 0 | 0 | 0 | 1 | 1 |
| $\mathbf{z} - \mathbf{c}$ | | $-1$ | $-1$ | $-1$ | 0 | 0 | 0 | 0 |
| | $\mathbf{a}_s$ | $\frac{2}{3}$ | $\frac{1}{2}$ | 1 | $\frac{1}{2}$ | 0 | 0 | $\frac{1}{2}$ |
| | $\mathbf{e}_2$ | $\frac{4}{3}$ | $\frac{1}{2}$ | 0 | $-\frac{1}{2}$ | 1 | 0 | $\frac{1}{2}$ |
| | $\mathbf{e}_3$ | $\frac{4}{3}$ | $\boxed{2}$ | 0 | 0 | 0 | 1 | 1 |
| $\mathbf{z} - \mathbf{c}$ | | $-\frac{1}{3}$ | $-\frac{1}{2}$ | 0 | $\frac{1}{2}$ | 0 | 0 | $\frac{1}{2}$ |
| | $\mathbf{a}_3$ | $\frac{1}{3}$ | | | | | | $\frac{1}{4}$ |
| | $\mathbf{e}_2$ | 1 | | | | | | $\frac{1}{4}$ |
| | $\mathbf{a}_2$ | $\frac{2}{3}$ | | | | | | $\frac{1}{2}$ |
| $\mathbf{z} - \mathbf{c}$ | | 0 | 0 | 0 | $\frac{1}{2}$ | 0 | $\frac{1}{4}$ | $\frac{3}{4}$ |

From the simplex tableau, $\mathbf{y}_0 = \left(0, \frac{1}{2}, \frac{1}{4}\right)'$ is an optimal vector, $z_0 = \frac{3}{4}$ is the value and $\mathbf{x}_0 = \left(\frac{1}{2}, 0, \frac{1}{4}\right)'$ is the optimal vector for the dual problem, which is found in the bottom row under the $e_i$'s. The corresponding optimal strategies for $P_1$ and $P_2$ are $\bar{\mathbf{x}} = z_0^{-1} \mathbf{x}_0 = \left(\dfrac{2}{3}, 0, \dfrac{1}{3}\right)'$, $\bar{\mathbf{y}} = z_0^{-1} \mathbf{y}_0 = \left(0, \dfrac{2}{3}, \dfrac{1}{3}\right)'$ and the value of the game is $\omega = z_0^{-1} - 1 = \dfrac{1}{3}$.

Since $z_1 - c_1 = 0$, another optimal vector can be obtained by replacing $e_2$ in the basis by $\mathbf{a}_1$. This gives another optimal strategy, $\mathbf{y}^* = \left(\dfrac{1}{3}, \dfrac{4}{9}, \dfrac{2}{9}\right)'$, for $P_2$. It can be proved that the set of optimal strategies for a player is a convex polytope (see exercise 9 on page 82), and in this example $\langle \bar{\mathbf{y}}, \mathbf{y}^* \rangle$ is the set of optimal strategies for $P_2$. The optimal strategy for $P_1$ is unique as the optimal vector for the linear programme is non-degenerate (see exercise 4 on page 66).

We conclude this chapter by summarising a procedure for obtaining a solution to a matrix game.

(i) Test for a saddle point, if one exists then a solution is immediate, if not proceed to (ii).

(ii) Eliminate dominated strategies.

(iii) If reduced matrix is $2 \times 2$, solve by the method given on page 76. If not, add a suitable constant $k$ to the elements of $A$ if the maximin is not positive, and then solve the related linear programme,

$$\begin{aligned} \text{maximise} \qquad & \mathbf{v}'\mathbf{y} \\ \text{subject to} \qquad & A\mathbf{y} + \mathbf{w} = \mathbf{u}, \quad \mathbf{y} \geq \mathbf{0}, \quad \mathbf{w} \geq \mathbf{0}, \end{aligned}$$

by the simplex method.

(iv) A solution to the game is given by $\bar{\mathbf{x}} = z_0^{-1} \mathbf{x}_0$, $\bar{\mathbf{y}} = z_0^{-1} \mathbf{y}_0$, $\omega = z_0^{-1} - k$, where $\mathbf{y}_0$ and $\mathbf{x}_0$ are optimal vectors for the linear programme and its dual, respectively, and $z_0$ is the value of the programme.

To determine the sets of optimal strategies for each player, the unreduced payoff matrix must be considered, as some optimal strategies may be removed by eliminating the dominated strategies. For further theorems on the structure of matrix games, the reader is referred to chapter 7 of the book by Gale.

## EXERCISES ON CHAPTER FIVE

1. Determine the payoff matrix for the following game. $P_2$, the bowler, can bowl a fast ball and a slow ball. $P_1$, the batter, can either hit or miss. If $P_1$ misses a fast ball or hits a slow ball, he is out and scores no runs. If $P_1$ hits a fast ball, he scores four runs and if he misses a slow ball, the game is played again and if he again misses a slow ball, he gets one run. ($P_1$ cannot recognise $P_2$'s ball in advance.)

2. Evaluate the maximin and minimax for the game $G = (S, T; \phi)$, where $\phi(s,t) \equiv 2st - s^2 + t^2 + 2s$ and $S \equiv T$ is the field of real numbers, and hence solve the game.

3. Solve the game with the $n \times n$ payoff matrix, $A = (a_{ij})$, where $a_{ij} = 0$ for $i \neq j$ and $a_{ii} > 0$ for $i = 1, 2, \ldots, n$. Use your result to solve the game of exercise 1.

4. A game, $G = (S, T; \phi)$ is called symmetric if $S \equiv T$ and $\phi(s,t) = -\phi(t,s)$ for all $s \in S$ and $t \in T$. Show that $\overline{x}(s)$ is an optimal strategy for $P_1$ if and only if $\phi(\overline{x},s) \geq 0$ for all $s \in S$.

Two aircraft each carry one rocket and the probability of the pilot of either aircraft hitting the other aircraft by firing his rocket from a distance $d$, is $e^{-d}$. What is the best distance for the pilots to fire from, given that each knows when the other has fired?

5. Solve the following game. $P_1$ is dealt one card and $P_2$ is dealt two cards from a pack of four aces. After looking at his card, $P_1$ guesses which ace remains in the pack and then $P_2$, having heard $P_1$'s guess and having inspected his own cards, also guesses which ace remains in the pack. If either player guesses correctly, he wins one from his opponent.

6. Use the simplex method to solve the game with payoff matrix

$$A = \begin{pmatrix} -3 & 3 & 2 \\ 5 & -1 & 0 \\ 0 & -2 & 3 \end{pmatrix}.$$

7. Commander $A$ has three aircraft with which to attack a target and his opponent $B$ has four anti-aircraft missiles to defend it. To destroy the target, it is sufficient for one of $A$'s aircraft to penetrate $B$'s defences. There are three corridors through which $A$'s aircraft can approach the target; $B$ can defend each corridor by placing any number of missiles in it and each missile is guaranteed to destroy one and only one aircraft. Find optimal strategies for $A$ and $B$.

8. Show that the linear programme,

maximise $\omega$

subject to $\mathbf{x}'A \geq \omega\mathbf{v}'$, $\mathbf{x}'\mathbf{u} = 1$ and $\mathbf{x} \geq \mathbf{0}$,

and its dual are feasible. Hence prove the minimax theorem for matrix games. ($\mathbf{u} = (1, 1, \ldots, 1)'_m$, $\mathbf{v} = (1, 1, \ldots, 1)'_n$)

9. Use the result of exercse 6 on page 12 to show that any non-negative solution, $\mathbf{y} \geq \mathbf{0}$, $\mathbf{w} \geq \mathbf{0}$, of the equations

$$A\mathbf{y} + \mathbf{w} = \omega\mathbf{u}, \quad \mathbf{v}'\mathbf{y} = 1,$$

can be expressed as a convex combination of the basic non-negative solution of the equations. Hence prove that the set of optimal strategies for $P_2$ is a convex polytope. ($\mathbf{u} = (1, 1, \ldots, 1)'_m$, $\mathbf{v} = (1, 1, \ldots, 1)'_n$)

# SUGGESTIONS FOR FURTHER READING

BEALE, E. M. L. *Mathematical Programming in Practice*, Pitman, London, 1968.

GALE, D. *The Theory of Linear Economic Models*, McGraw Hill, New York, 1960.

KREKÓ, B. *Linear Programming*, Pitman, London, 1968.

MCKINSEY, J. C. C. *Introduction to the Theory of Games*, McGraw Hill, New York, 1952.

NEUMANN, J. von and MORGENSTERN, O. *Theory of Games and Economic Behaviour*, Princeton, 1944.

SIMMONARD, M. *Linear Programming*, Prentice Hall, 1966.

# SOLUTIONS TO EXERCISES

Chapter I

1. Convex, not convex, convex.
3. $C$ is empty ($\alpha > 2$), $C$ is convex polytope ($0 < \alpha \leq 2$), $C$ is half-line ($\alpha \leq 0$).

Chapter II

4. $\mathbf{x} = \left(0, \dfrac{8}{5}, \dfrac{19}{5}\right)'$, value $= -\dfrac{68}{5}$.

6. Minimise $\alpha_{n+1}$,

   subject to $\displaystyle\sum_{i=1}^{n} \alpha_i \phi_i(x_j) + \alpha_{n+1} \geq f(x_j)$ for $j = 1, 2, \ldots, m$,

   and $\displaystyle\sum_{i=1}^{n} \alpha_i \phi_i(x_j) - \alpha_{n+1} \leq f(x_j)$ for $j = 1, 2, \ldots, m$.

7. Dual: maximise $\mathbf{y}'\mathbf{b}$,

   subject to $\mathbf{y}'A \leq 0$, $\mathbf{y}'\mathbf{b} \leq 1$ and $\mathbf{y} \geq 0$.

10. Maximise $\mathbf{c}'\mathbf{x}$,

   subject to $A\mathbf{x} = \mathbf{b}$ and $\mathbf{l} \leq \mathbf{x} \leq \mathbf{u}$.

   Dual: minimise $\mathbf{y}'\mathbf{b} + \mathbf{w}'\mathbf{u} - \mathbf{z}'\mathbf{l}$,

   subject to $\mathbf{y}'A + \mathbf{w}' - \mathbf{z}' \geq \mathbf{c}'$ and $\mathbf{w} \geq 0$, $\mathbf{z} \geq 0$.

Chapter III

1. $x_{ij} = \begin{pmatrix} 1 & 2 & 3 \\ 4 & 0 & 0 \end{pmatrix}$, cost $= 34$.

2. Minimum cost $= 65(\lambda > 6)$, $= 59 + \lambda(4 < \lambda \leq 6)$, $= 47 + 4\lambda(0 \leq \lambda \leq 4)$.

3. Optimal policy for caterer is on Monday to buy 30 from shop, on Tuesday to buy 30 from shop and to use 10 from 1-day laundry, on Wednesday to use 20 from 2-day laundry and 40 from 1-day laundry, and on Thursday to use 40 from 1-day laundry.

4. $C_1$ does $J_4$, $C_2$ does $J_1$, $C_3$ does $J_2$ and $C_4$ does $J_3$.

   $C_1$ does $J_2$, $C_2$ does $J_1$, $C_3$ does $J_4$ and $C_4$ does $J_3$.

6. Equivalent transportation problem has supplies $s_i = s$ for $i = 1, \ldots, 4$, demands $d_i = d_i$ for $i = 1, \ldots, 4$ and $d_5 = 4s - d_1 - d_2 - d_3 - d_4$, and cost matrix $c_{ij}$, where $c_{i5} = 0$, $c_{ij} = \infty$ for $i > j$, and $c_{ij} = c_i + k(j - i)$ otherwise. For feasibility $s \geq d_1$, $2s \geq d_1 + d_2$, $3s \geq d_1 +$

$+ d_2 + d_3$ and $d_5 \geq 0$.

$$\text{Optimal } x_{ij} = \begin{pmatrix} 4 & 2 & 0 & 0 & 0 \\ 0 & 5 & 0 & 0 & 1 \\ 0 & 0 & 5 & 1 & 0 \\ 0 & 0 & 0 & 6 & 0 \end{pmatrix}, \text{ cost } = 106.$$

7. Use equilibrium theorem and assume that the result does not hold.

## Chapter IV

1. $\mathbf{x} = \left(0, \dfrac{16}{29}, 0, \dfrac{66}{29}, 0, \dfrac{11}{29}\right)'$, value $= -\dfrac{3}{29}$.

2. No optimal solution.

5. $\mathbf{x} = (1, 2, 0, 1)'$ is optimal for $\lambda \geq 3$ with value $= \lambda + 5$; no optimal solution for $\lambda < 3$.

6. $\mathbf{y} = \left(0, \dfrac{2}{3}, \dfrac{8}{3}\right)'$, value $= \dfrac{32}{3}$; optimal solution to dual is $\mathbf{x} = \left(\dfrac{1}{3}, \dfrac{2}{3}\right)'$.

7. $\mathbf{x} = \left(\dfrac{11}{2}, \dfrac{9}{4}, 7\right)'$, value $= 47$.

8. 25% $C_1$, 75% $C_4$, cost $= 3s\ 3d$. Reduce by $3d$, $33\frac{1}{3}$% $C_2$, $66\frac{2}{3}$% $C_4$.

## Chapter V

1.
$$\begin{pmatrix} 4 & 0 & 0 \\ 0 & 4 & 0 \\ 0 & 0 & 1 \end{pmatrix}$$

2. Maximin $=$ minimax $= \dfrac{1}{2}$, solution is $\left(s = \dfrac{1}{2}, t = -\dfrac{1}{2}; \omega = \dfrac{1}{2}\right)$.

3. Optimal strategies $\bar{\mathbf{x}} = \bar{\mathbf{y}} = \left(\dfrac{\omega}{a_{11}}, \dfrac{\omega}{a_{22}}, \ldots, \dfrac{\omega}{a_{nn}}\right)'$ and $\omega = \left(\sum_{i=1}^{n} a_{ii}^{-1}\right)^{-1}$. Solution to game of exercise 1 is $\bar{\mathbf{x}} = \bar{\mathbf{y}} = \left(\dfrac{1}{6}, \dfrac{1}{6}, \dfrac{2}{3}\right)'$, $\omega = \dfrac{2}{3}$.

4. Best distance is given by $e^{-d} = \dfrac{1}{2}$.

5. $\begin{pmatrix} 0 & -1 \\ -\frac{1}{3} & 0 \end{pmatrix}$, solution is $\bar{\mathbf{x}} = \left(\dfrac{1}{4}, \dfrac{3}{4}\right)'$, $\bar{\mathbf{y}} = \left(\dfrac{3}{4}, \dfrac{1}{4}\right)'$, $\omega = -\dfrac{1}{4}$.

6. $\bar{\mathbf{x}} = \left(\dfrac{1}{2}, \dfrac{1}{2}, 0\right)'$, $\bar{\mathbf{y}} = \left(\dfrac{1}{3}, \dfrac{2}{3}, 0\right)'$ or $\left(\dfrac{1}{4}, \dfrac{1}{4}, \dfrac{1}{2}\right)'$, $\omega = 1$.

7.

|  |  | B's strategies of deploying missiles | | | |
| --- | --- | --- | --- | --- | --- |
|  |  | (4, 0, 0) | (3, 1, 0) | (2, 2, 0) | (2, 1, 1) |
| A's | (3, 0, 0) | $\frac{2}{3}$ | $\frac{2}{3}$ | 1 | 1 |
| strategies | (2, 1, 0) | 1 | $\frac{5}{6}$ | $\frac{2}{3}$ | $\frac{2}{3}$ |
| for | (1, 1, 1) | 1 | 1 | 1 | 0 |
| aircraft |  |  |  |  |  |

Solution $\bar{\mathbf{x}} = (\frac{1}{3}, \frac{2}{3}, 0)'$, $\bar{\mathbf{y}} = (0, \frac{2}{3}, 0, \frac{1}{3})'$ (not unique), $\omega = \frac{7}{9}$.

86

# Index